THE BIOLOGY OF POPULATION GROWTH

The Biology of Population Growth

ROBERT L. SNYDER

CROOM HELM LONDON

First published 1976
©1976 by Robert L. Snyder

Croom Helm Ltd
2-10 St John's Road, London SW11

ISBN 0-85664-067-0

The illustrations are reproduced from drawings by Ellen E. Bowler,
Staff Artist at the Philadelphia Zoological Garden

Printed in Great Britain
by Biddles Ltd, Guildford, Surrey

CONTENTS

Tables
Figures

TABLES

FIGURES

1 ECOLOGY AND POPULATION

The purpose of this book is to present a contemporary view of an important ecological principle — that living organisms are integrated groups or populations of individuals dependent upon one another and upon elements of the environment for their existence and perpetuation. The science of ecology is the study of the relationships between living organisms and their environment. One aspect of this science treats living organisms relevant to their physical environment, which is merely the study of the physical conditions required for life on this earth. A second aspect of ecology is concerned with an individual organism's relationship to other living matter in its universe. Some of the more obvious biotic or biological relationships are represented by forms of parasitism and predation, such interactions being between organisms unrelated genetically. The most complex and possibly the most important biotic relationships exist between related organisms or organisms of the same species. The treatment of genetically related organisms as integrated groups or evolved aggregations of interacting organisms simplifies conceptualization and leads to applicable laws of ecology or at least workable hypotheses. At the very least, this way of looking at living organisms is a convenient way of organizing pertinent information about reproduction, mortality, and animal behavior.

Today the mass communications media tend to use a single word to symbolize each important complex problem pressuring mankind. Thus, 'energy crisis' covers the entire problem of utilizing fossil fuels without destroying ourselves in the process. 'Ecology' stands for the problem of the wholesale destruction of wildlife habitat that accompanies technological and social progress. And 'population' represents a vague uneasiness about the possibility of there being too many people in the world. The average person, understandably confused, is caused to wonder whether the problems are really as bad as portrayed.

The population problem is the underlying theme of this book. The array of complex ecological interactions which determine the numbers of animals and plants that exist in a given area through time should eventually be understood in terms of the basic principles governing these interactions. Unfortunately, population biology is still in its developmental stages. Several principles of population dynamics have been proposed, tested in the laboratory, and studied in the field, but ecologists cannot agree on their validity or applicability. There are two opposing schools of thought about what controls population growth. One school believes that extrinsic agents such as food supply, weather, predators, and disease are responsible for regulating population size. The other looks to intrinsic mechanisms, inherent characteristics of the population members collectively.

We humans occupy a unique position in the ecological world. Down through the ages we have had a profound effect on populations of plants and animals, having caused the extinction of many species and the extirpation of others from vast regions. In a word, we have played havoc with the natural scheme of things.

Man may have reached the stage in history when he can exercise some control over populations — perhaps even his own. Therefore, it behooves us all to attempt to understand the principles of ecology in general and those of population in particular. What about the disagreements between the ecologists themselves? Arguments about details, particulars, and 'laws' should not be an insurmountable obstacle to the study of populations. History has taught us that conflict and disagreement are inevitable in scientific fields. What was dogma yesterday may be disproven tomorrow. What appears to have been two unalterable opposing schools has a way of blending in time. It often happens that both sides were partly right. Ultimately, the readers must judge for themselves on the basis of the evidence presented. Such is the case with this book.

Both plants and animals make up the living universe and it is impossible to treat one without the other. However, the animal kingdom is infinitely more complex organismically, and many of its members have unique properties such as the power of locomotion and a highly developed nervous system which enable them to respond to environmental exigencies in ways impossible for members of the plant kingdom. Therefore, in this book plants will have to take a back seat. The emphasis will be on animal populations.

Population Problems

Assessment of environmental impact requires certain judgments about the values of the organisms affected. If these were entirely monetary, the problem would be simple. A 16-ounce can of salmon in 1975 sold from $.89 to $3.65 depending on the species. Thus, a ten-pound Atlantic salmon *(Salmo salar)* might have been valued at approximately $30.00 based on its food value. However, in 1975 the province of New

Brunswick in Canada figured that each salmon caught by a non-resident angler was worth many hundreds of dollars to its residents. In the same sense, earthworms might sell for 25 cents a dozen for fish bait, but their value in an ecosystem which depends on these invertebrates to break down humus into chemical forms used as nutrients by plant species is inestimable. Ultimate evaluation in any case depends on how many and what kinds of plants and animals are involved. To handle this kind of problem, biologists have evolved the principle of population.

Population

The word 'population' is derived from 'populus' meaning people. To sociologists a population is the total number of persons inhabiting a country, city, district, or area. Biologists evidently borrowed the term to define the total number of organisms inhabiting an area or region. Statisticians include both animate and inanimate units in their populations. Their precise definition is any finite or infinite aggregation of individuals, not necessarily animate, subject to statistical study.

We are indebted to the American statistician Raymond Pearl for developing a precise definition of population for ecology:

> A population is a group of living individuals set in a frame that is limited and defined in respect of both time and space. The biology of populations is consequently a division or department of group biology in general. The essential and differentiating feature of group biology is that it considers groups as wholes. It aims to describe the attributes and behavior of a group as such, that is an entity in itself, and not as the simple sum of the separate attributes and behaviour of the single individual organisms that together make up the groups.

The barnacles on the shell of the horseshoe crab in Figure 1 can be considered collectively as a population.

Biosphere

What is now recognized as population biology was not created in a fortnight, but evolved as a natural consequence of the development of ecology as a science. Animal and plant life occupies the whole surface of the earth including land and sea. The whole space occupied by living organisms is called the biosphere. This thin film of air, water, and soil is roughly ten miles deep, or one four-hundredth of the earth's radius. Ecosphere is another word for the same space. Plant geographers

Figure 1. A Population of Barnacles on the Shell of a Horseshoe Crab *(Limulus* sp.)

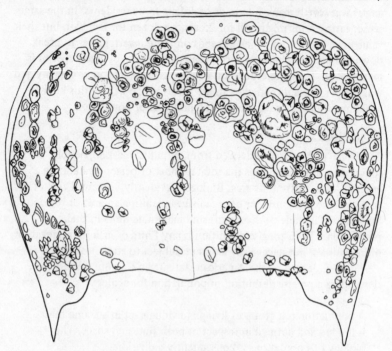

were the first to give serious attention to studies of the biosphere in an ecological context. They were concerned with the numbers and forms of plant life represented in various habitat conditions. Thus, the botanists had a classification and nomenclature worked out for the major divisions of the biosphere before zoologists were much interested in the problem. Although animal ecologists naturally borrowed much of their terminology from the phytogeographic system, they did make some revisions because of the inherent differences between plants and animals. Also, as with most classification schemes, it should be admitted that no unanimity exists with respect to both fundamental subdivisions and terms. However, this is not an insurmountable problem as long as precise definitions are employed. Population students can use the terms interchangeably as long as the meanings are clear.

Niche

The primary topographic unit is the biotope or niche. Such a unit has been described as an area showing uniformity in the principal habitat conditions, and in current usage has come to mean the place where a species lives. Discussions of niches usually imply a degree of exclusiveness in which the microhabitat, the smallest possible subdivision that insures uniformity of conditions met nowhere else, is occupied by a single species. Actually, the originators of the term 'ecological niche' were not nearly so precise. Joseph Grinnell defined a niche as the entire range of physical environments that a species was found in. Charles Elton envisioned it as an animal's place in the biotic environment. His conceptualization was eloquent: 'It is therefore convenient to have some term to describe the status of an animal in its community, to indicate what it is *doing,* and not merely what it looks like, and the term used is "niche".'

Naturally, such a fundamental unit in the classification scheme of ecology would become the subject of much theoretical discussion. American ecologists G. E. Hutchinson and Robert MacArthur, for example, have developed the concept of the multidimensional fundamental niche. In this, the position of a population is defined in relation to as many axes as an investigator can measure. According to MacArthur, there is no limit to these, hence the fundamental niche has an infinity of dimensions.

The biotope as an ecological subdivision of the biosphere is as basic as the concept of the species is in the systematic classification of living things. And as species are combined into genera and these into families and orders, the biotopes can be grouped according to their resemblances into biochores.

The biochores and superbiochores are finally combined into still higher groupings which are called biocycles by some and ecosystems by others. In America the ecosystem has attained common usage. The biosphere can be divided into three fundamentally distinct ecosystems: ocean, fresh water, and land. Each ecosystem contains animal species that are almost exclusively adapted to their own peculiar habitat conditions. Certain species live in more than one ecosystem at different times during their life cycles, for example, salmon, eels, and shad occur in the ocean and in fresh water; amphibia, in fresh water and on land; and some birds in all three, but they are so much a minority that they prove the principle of exclusion.

Ecosystems have their respective fauna and flora called biota when both plants and animals are included in one category. The plants considered with regard to their ecological relations rather than their taxonomic affinities are called the phytome and the animal life, similarly considered the zoöme.[1]

Communities

Historically, both zoologists and botanists have recognized so-called natural assemblages of organisms which are typical of certain habitat conditions. The birch-beech-maple forests of north-central Pennsylvania, for example, have indicator species of certain trees, shrubs, grasses, etc. typically found in nearly every part of the biochore. Moreover, such forests have typically certain animal species, black bears *(Ursus americanus)*, gray squirrels, *(Sciurus carolinensis),* great horned owls *(Bubo virginianus)*, porcupines *(Erethizon dorsatum)* and bobcats *(Lynx rufus)*, to name the most important. Other animals, the white-tailed deer *(Odocoileus virginianus)* and wild turkey *(Maleagris gallapavo)*, for example, are plentiful there today, but such species are not usually considered type species because they are widespread in other forest associations as well. The plants and animals are so characteristic of the great area covered by the dominant birch-beech-maple vegetation that to find an atypical form would be reason enough to suspect a significant local variation in soil fertility, water table, or hydrogen ion concentration (pH).

The natural groupings of plant and animal species within certain easily discernible habitat conditions led to the formulation of the community concept. The major community is defined as a natural assemblage of organisms which, together with its habitat, has reached a survival level such that it is relatively independent of adjacent assemblages of equal rank; to this extent, given radiant energy, it is self-sustaining.[2] Thus, in the birch-beech-maple forest the ecologist would find that the phytome was dependent upon the soil nutrients, porosity of the soil, the subsurface characteristics, rainfall, topography, etc. and that the animal species were adapted to utilize the birch-beech-maple forest for food and shelter. Certain predacious species associated with the herbivores complete this community.

The community concept is considered an essential element of modern ecology. The community is a primary study unit. By defining the overall self-sustaining subdivision of the ecosystem, the ecologist is able to separate its interrelated parts to determine how the whole system operates.

The Gene Theory

Living organisms are physicochemical structures capable of utilizing solar energy, directly or indirectly, to grow and multiply; some achieve mobility or at least movement of certain parts. Organisms exist in many forms, from the simplest virus particle to the complex multicellular plant or animal. What separates living from non-living entities is basically the ability for replication possessed by the former. Some may argue that certain molecules containing carbon may replicate under special conditions, but this does not bother biologists who, unlike laymen, know no lines of demarcation separating living from non-living material, nor for that matter plant life from animal life. Such differentiations are surely conceptual and do not correspond to reality.

If life is the property of replication, then what of the physical and chemical nature of the genetic material and how does it work? We speak of genes as the determinants of heredity and populations as natural assemblages of organisms with the same heredity; therefore, at least a rudimentary understanding of current gene theories is desirable.

Although the full story is not yet told, it can be stated with some assurance that the substance most directly associated with the storage and perpetuation of genetic information is the deoxyribonucleic acid (DNA) in the chromosomes of the germ cells. DNA can be broken up by acid or enzymatic hydrolysis in successive stages as follows: DNA → nucleotides → nucleosides + phosphoric acid → purine and pyrimidine bases + deoxyribose. In most samples of DNA, four heterocyclic bases predominate — the purines adenine (A) and guanine (G) and the pyrimidines thymine (T) and cytosine (C) (Fig. 2).

Figure 2. The Four Heterocylic Bases of DNA

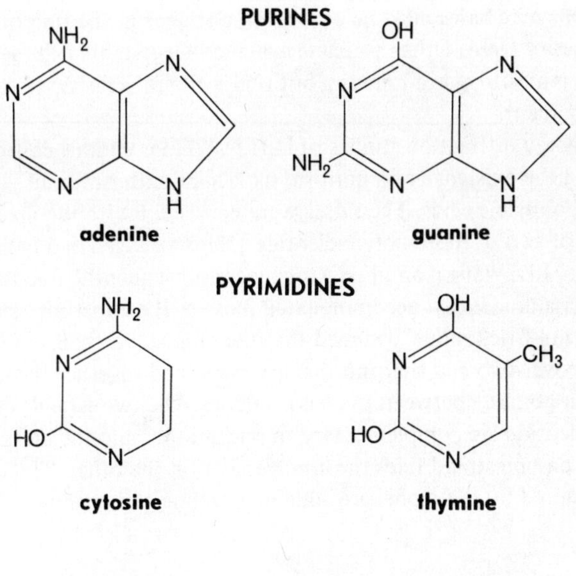

PURINES

adenine   guanine

PYRIMIDINES

cytosine   thymine

Biochemists, as a result of a series of very skillful experiments and deductions utilizing chromatographic techniques and X-ray diffraction and titration studies, have been able to visualize component parts of the DNA molecule and the three-dimensional aspects of its surface.

The sugar, deoxyribose (Fig. 3) together with phosphate residues forms the chemical backbone of the DNA molecule. Deoxyribose has three hydroxyl (OH) groups. The one at carbon 1 is bound to one of the four heterocyclic bases. Those at carbons 3 and 5 are linked through

Figure 3. Deoxyribose

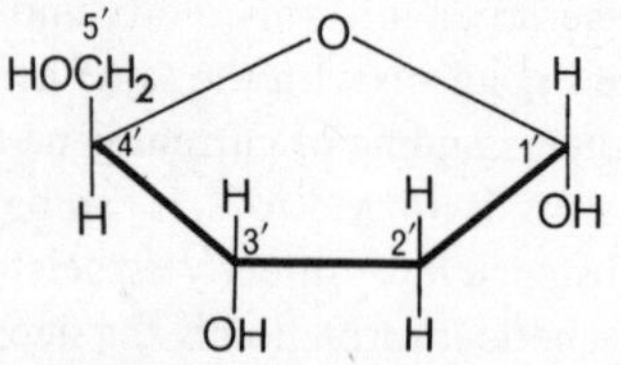

2-Deoxy-D-ribose

phosphate diester bridges to join other deoxyribose molecules (Fig. 4).

Rounding out the purely chemical nature of the DNA structure requires only the identification of the purine or pyrimidine bases at the various carbon 1 positions. But this of course is where complexity rears its ugly head. Evidence is strong that DNA molecules are constructed according to a meaningful plan and that the arrangement of heterocyclic bases must be of great importance in the functional properties of DNA. Other available evidence suggests that the sequence of bases is anything but random, but this is about as far as we can go at the moment.

The X-ray diffraction studies of M.H.F. Wilkins and his colleagues indicated the existence of a uniform molecular pattern for all deoxyribonucleic acids. Their data were consistent with the presence in DNA of two or more polynucleotide chains arranged in a helical structure. J.D. Watson and F.H. Crick were subsequently able to construct a model that accommodated most of the experimental facts. The Watson-Crick model assumed that the number of chains in the molecule was two and that the chains were joined together through hydrogen bonding between the base residues. The two strands of polynucleotide are complementary to one another and the arrangement of bases on one strand fixes the arrangement on the other. The base pairs required by the model are adenine-thymine and guanine-cytosine.

Thus, if the bases along one strand are arranged in the order
G-G-T-C-A-A, the opposite bases on the complementary strand would
be C-C-A-G-T-T-.

The Watson-Crick model represents DNA as a double helix. If the
DNA in a single haploid cell were present as a single double helix, it

Figure 4. The Nature of the Linkage Between Individual Deoxyribonucleotides
in Deoxyribonucleic Acid

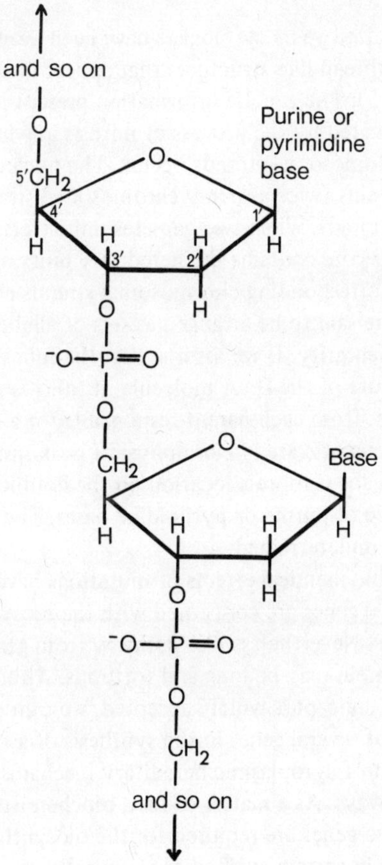

could be as long as 15 metres. Such a coiled structure would make 4.4×10^9 turns around the screw axis. It is unlikely that the double helix runs the full length of the chromosome, however, so models accommodating a multiple of double helices in each chromosome have been proposed.

Until quite recently the term 'gene' was employed to convey the purely abstract concept of a unit of heredity. It represented a quantum of specific information that in some way controlled the bio-synthesis of cellular chemical structures, or in more cautious terms, of some functional units.

For over a hundred years cytologists have been aware of chromosomes as visible rods or thread-like structures that appear in the nucleus during cell division (Fig. 5). The genetic information present in a cell is accurately perpetuated by the process of mitosis in which exact duplication of chromosomal strands occurs. The nucleus of a somatic cell (diploid) contains twice as many chromosomal strands as a germ cell or gamete (haploid). When two gametes unite (fertilization), the resulting diploid *zygote* contains the hereditary units of both parents arranged so that corresponding chromosomal strands are paired. The hereditary units are said to be arranged as sets of allelic genes in exact physical complementarity. If we assume that the functional genetic unit is a substructure of the DNA molecule, it follows that the parts of the double helix from each parent responsible for a specific phenotypic characteristic are in close physical proximity in the diploid cell. Therefore, a gene is more a location on the double helix than any particular sequence of purine or pyrimidine bases. The ultrastructure of the gene is still undetermined.

Studies of the biochemical effects of mutations have supported the idea that individual genes are concerned with the biosynthesis of individual proteins. Nevertheless, the pathway from gene structure to phenotypic protein may be long and tortuous. Although the 'one gene-one enzyme' concept is widely accepted, we cannot rule out the combined action of several genes in the synthesis of a single protein or the involvement of cytoplasmic hereditary mechanisms in biosynthetic pathways. As a matter of fact, biochemists have discovered that two genes are required for the biosynthesis of a single polypeptide chain in certain antibodies.[3] This discovery may well have significant implications in our understanding of group versus individual selection during the process of evolution and will be alluded to again in Chapter 6.

Figure 5. Artist's Drawing of Chromosomes From a California Sea Lion *(Zalophus californianus)*

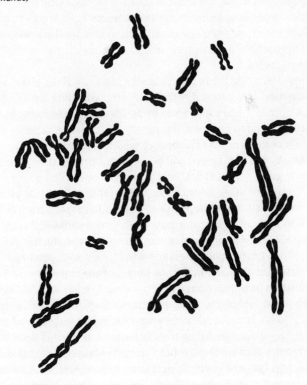

The Species Concept

The ultrastructure of the gene is yet unknown and the genetic code remains 'unbroken', but this is no great hindrance to our concept of species. To appreciate the meaning of species we must explain how they are made, i.e. how new species arise from old. As the genes are modified (mutation or change in the sequence of bases in the DNA or other structural modification) and reshuffled, occasional unique organisms will appear within restricted populations with traits that enable them to explore niches that were unattainable by their predecessors. Given time the summation of new characteristics results in an aggregation of interbreeding organisms fully at home in a new environment. This population of unique individuals will only become a new species when cross-fertilization between them and their distant ancestors is no longer possible.

Isolating Mechanisms

Divergent evolution, the branching of the phylogenetic tree to form new species, depends upon reproductive isolation. Isolating mechanisms prevent gene flow from one population to another. Isolation of closely related groups of animals from one another is essential to the formation of new species.

Initial isolating factors are probably more often the effect of geographical or ecological barriers to gene flow. Once the animals of a single population are split apart by an impenetrable geographic screen, the two separated groups will gradually drift apart genetically through random changes in gene frequency, genetic fixation, and mutation. Eventually the two groups will be separated permanently through physiological, psychological, and genetic incompatibility.

In some instances distance alone may act as an isolating mechanism in a widely distributed population. Genes filter slowly from one location to another, and this may result in morphological variations with no sharp lines of demarcation. Complete reproductive isolation as a result of mere spatial isolation would seem to depend mainly on population density and individual ranges of the animals.

Isolation of populations by topographic barriers is considered one of the prime factors inducing speciation. Examples of such barriers are wide streams, canyons, deserts, waterfalls, ridges, valleys, and mountains.

Sexual isolation resulting from behavioral or psychological interference with mating has been termed ethological isolation.[4] The distinction between psychological and physiological isolation may be unwarranted now because behavior, especially sexual behavior, is intricately involved with the endocrine organs. Generally, however, the evolution of different courtship patterns, different chemical sex attractants, and different call notes, to name a few possibilities, cements the isolation of populations initially separated geographically or spatially.

The Species and the Population

Once the species concept is accepted, it is evident that the population is a fundamental ecological unit and an elemental part of the biotic community. The species is conceptually an infinite population of interbreeding organisms. When we speak of the southern woodchuck *(Marmota monax)* as a species, the term is employed collectively to include all of the *Marmota monax* in the world. If one were to study the entire population of *Marmota monax* existing during a specified period of time, it would be necessary to determine the limits of their

distribution and draw the study area to encompass all woodchucks
capable of interbreeding to produce viable offspring. Naturally, this
would be impossible, hence small, finite subunits of the universal
population are selected instead. The choice of boundaries for the study
of populations and the time frame required are important, as data
collected are considered representative of the species *Marmota monax*.
To help with this sort of problem ecologists utilize statistical theory.
Sampling techniques and sampling theory determine the sample size
or number of animals that should be studied. Statistical methods
furthermore indicate how representative of the species the data
collected are likely to be.

The Genetic Basis for Speciation

The definition of population (species) emphasizes likeness of
individuals, but this concept should not be taken for granted.
Individuals of the same species share the same genetic material in the
autosomal chromosomes, except that males and females have different
sex chromosomes. The Y chromosome of mammals is usually smaller
than its complementary mate, the X chromosome. The heterozygous
individual has an X and a Y chromosome in each somatic cell and
develops as a male. The female, the homozygous individual, possesses
two X chromosomes in each somatic cell. In birds, the reverse is true,
the female being the heterozygous and the male the homozygous
individual. The sex chromosomes of birds are called Z and W, the
ZZ combination being the male. Occasionally, the Y or the W
chromosome as the case may be is entirely missing in some species.

The karyotype (the number, size, and shape of the chromosomes)
is the same for every male and for every female member of the
species. The sameness of the chromosome arrangement in the various
individuals makes inbreeding possible. However, similarity of karyotypes
is not a prerequisite for species designation. All living members of the
family Camelidae have the same karyotype but there are six distinct
species recognized.[5] The llama *(Lama glama)* and the bactrian camel
(Camelus bactrianus) are distinctly different phenotypically but have
the same karyotype. Thus, speciation and acquisition of isolating
mechanisms are apparently often entirely genic in nature, not involving
any visible karyological change.

Sexual dimorphism is responsible for most of the variability
among individuals in a population, but there are variables unrelated to
sex. In Cameron County in northern Pennsylvania, roughly half the
'gray squirrels' observed are black instead of gray. The early settlers

distinguished two species of tree squirrels, the black squirrels inhabiting the upper New England states and Canada and the gray tree squirrels inhabiting an area from New York State southward. Parts of Pennsylvania were considered the mixing grounds where both species abided. Of course, the black squirrel is a melanistic phase of the common eastern gray squirrel and blacks and grays are often found in the same litter. Selective pressures evidently favor the black coloration in northern environs, as the 'grays' in Canada are nearly all black. The reverse is true in southern Pennsylvania and points even further south because in these regions a black squirrel is a curiosity.

Scotch grouse *(Lagopus scotius)* in Great Britain show variations in the shades of red, while ruffed grouse *(Bonasa umbellus)* in North America occur in red and gray (silver) phases. The young of the year in a flock of whistling swans *(Cygnus columbianus)* stopping to rest during the fall migration on the Conemaugh River in Pennsylvania are dull gray compared to their snow-white parents. Besides the differences in coloration, the individuals in a population commonly come in a variety of sizes and shapes, not always due to differences in ages. This kind of individual variability is carried to extremes in the insects and in some other invertebrate species as well. It should be apparent by now that the basic similarity that we speak of is not physiognomical but genetic. The one characteristic that allows collectivity is the ability of individuals to breed with one another to produce viable offspring which in turn interbreed to preserve the gene pool of the species.

Population Structure

Once the matter of individuality is set aside, it follows naturally that questions about the relations of these individuals to one another are of paramount importance. How many animals are there in the population? What is the sex ratio? How many young are produced each year? How are the animals distributed spatially? How many animals die each year? These and a number of other related questions are grist for the mill of population biology.

Density

Density or the number of organisms per unit of area or volume is perhaps the most interesting characteristic of a population and stimulates more than its share of theorizing. A characteristic mean level of density is fairly typical of a particular species. In general, the smallest animals have the highest densities. Insects, which are estimated to account for nearly three-fourths of all the species in the animal kingdom, are the most abundant and widespread of the land animals, being the principal invertebrates that can live in dry environments and the only ones able to fly. The sea is not their favorite habitat, there being few submerged marine forms and only water striders on the ocean surface. In their most favorite habitats insects reach densities of seven million per acre. Earthworms number 800,000 per acre in fertile soil. Single-celled protozoa are counted by the millions in volumes of one or two litres.

The vertebrate animals, ranging in size from tiny cyclostome larvae only 12 to 15mm long and shrews weighing 2.3 grams to the 90-foot-long (27.4 meters) blue whale *(Balaenoptera musculus)*, the largest creature that has ever lived, occur at low levels of density comparatively, but still total numbers of the large birds and mammals reach astonishing figures. Passenger pigeons *(Ectopistes migratorius)* in North America were once so numerous as to defy imagination. Stories are told of vast flights of these birds shutting out the sun during the day, and so many birds lighting on tree branches at one time that the boughs broke under their weight. John Audubon, the American naturalist, in 1813 observed one flock between Hardensburgh, Ohio, and Louisville, Kentucky, that he estimated at 1,115,000,000. It was believed this flock alone would consume 8,712,000 bushels of food a day. The American bison *(Bison bison)* once roamed the plains of North America by the million and today are reduced to scattered free-ranging herds numbering into the hundreds. The stories told about the colossal numbers of birds and mammals in Africa are classic. However, to the average ecologist in America and Europe these accounts are only figures in books and journals. For this reason, one is liable to ignore the large populations of animals that are still living in close proximity to the large metropolitan areas of the highly industrialized nations.

The Philadelphia Zoological Garden is confined to an area of 42 acres within the city of Philadelphia. The population of capitve animals varies, but an average 500 mammals, 1,500 to 2,000 birds, and 200 to 500 reptiles and amphibians is typical. In 1969 and 1970 I counted 2,500 ± 300 pigeons, *(Columba livia)* the native rock doves of England, an introduced species in America, feeding on the zoo property each day. Each evening vast flocks of English starlings *(Sturnus vulgaris)*, also aliens, light in the large trees of the Zoo, preparatory to going to roost under the neighboring Girard Avenue Bridge that crosses the Schuylkill River nearby. Of course, without elaborate sampling techniques, it was impossible to determine the total population using the bridge superstructure for a roosting site, but

100,000 or even 200,000 would not be considered an unlikely estimate. Living in the Zoo grounds are several more free-ranging species which are not behind bars and which obtain food by their own efforts. The list of mammals includes the eastern gray squirrel, the cottontail rabbit *(Sylvilagus floridanus)*, the raccoon *(Procyon lotor)*, an occasional opossum *(Didelphis virginiana)*, the Norway rat *(Rattus norvegicus)*, the brown house mouse *(Mus musculus)*, the meadow vole *(Microtus pennsylvanicus)*, and the deer mouse *(Peromyscus maniculatus)*. If this survey were expanded to include the adjacent Fairmount Park, the white-tailed deer could be added to the list.

One is conditioned to believe that wild creatures were formerly abundant and that man's activities have extirpated the vast majority of species or reduced their numbers to mere remnants of their former magnitude. While it is true that many species were destroyed and that others were greatly reduced in number, some species, like the southern woodchuck, benefited from the changes in habitat and are now existing at populations unheard of before the white man settled in North America. The woodchuck was probably at home in the forest clearings and the meadows created by beaver dams, but as the early settlers cleared the land for farming they provided vast new habitats for these herbivorous mammals. The cottontail rabbit also benefited from these changes. White-tailed deer are not so plentiful in the climax forest, but thrive in second growth forests where browse is plentiful and within easy reach. Pennsylvania was largely covered by a climax forest of white pine *(Pinus strobus)* and hemlock *(Tsuga sp.)* in 1700. The population density of the white-tails in such a forest was very low, and the animals distributed near the meadows produced by beaver ponds or in the shrub thickets growing up in locales where natural fires and windstorms had brought down the forest giants and opened the soil to sunlight. While at first the deer were nearly extirpated by uncontrolled hunting and clear cutting of the forest, they soon responded to the abundant food supply of the second growth forests. From 1915 to 1974 nearly 3,572,225 white-tailed deer were harvested by regulated hunting and today the deer herd is estimated at 600,000 (November 1974). The history of the population dynamics of white-tail deer in Pennsylvania illustrates the important relationship between numbers and food supply.

Biomass

Ecologists use biomass to describe the living matter in a

given habitat. Biomass is another measure of population size and is expressed either as the weight of organisms per unit area or as the volume of organisms per unit volume of habitat. Biomass and density together give the physical dimensions of a population.

Energy Cycles

The necessity for finding sufficient quantities of the right kind of food is the primary driving force of all animals. They have to depend ultimately upon the plant kingdom for their supplies of energy, since plants alone are able to utilize the energy from the sun to combine simple chemicals into large molecules which are the building blocks of life. The green leaf pigment chlorophyll is the magic catalyst that allows plants to capture solar energy. This process of photosynthesis combines carbon dioxide and water into molecules of carbohydrate and releases oxygen into the atmosphere.

Animals are able to use (digest) plant materials as a source of chemical compounds for the manufacture of skin, bones, and muscles. Parts of the carbohydrate molecules are burned to provide energy for locomotion and physiological functions. The combustion of these molecules in animal cells is termed respiration. The process consumes oxygen from the atmosphere, creates heat, and returns carbon dioxide and water to the atmosphere.

According to the First Law of Thermodynamics, energy can neither be created nor destroyed, although it may be changed from one form to another. For example, in photosynthesis light energy is changed to a new form of energy in the chemical bonds of the carbohydrate molecules. The Second Law of Thermodynamics states that a loss of usable energy will occur in any transfer of energy. In essence, a certain amount of the energy will be degraded from an available, concentrated form to an unavailable, dispersed form. In the photosynthetic system, normally one percent or less of the sunlight falling on green plants is converted to the chemical-bond energy that is available to animals eating the plants. Roughly ten percent of this store of energy in the plants may turn up as available energy in the chemical bonds of the plant eaters. And roughly ten percent of that energy may in turn be incorporated into the chemical bonds of animals that eat these animal plant eaters. Thus, carnivores may capture about one ten-thousanths ($.01 \times .10 \times .10$) of the solar energy impinging on the green plants in an area.

Food Chains

Herbivorous animals derive their food needs directly and solely from plant material. Carnivores get their energy from the sun at third hand by catching and eating herbivores of assorted sizes. Omnivores eat both plants and animals. In general, the smaller animals occur in large numbers. For example, a large number of insect species feed directly upon plant material, and these occur in prodigious numbers. Several predacious insects prey upon these plant eaters; insect predators are less numerous than their prey. Small mammals such as mice, insectivores, rats, and muskrats are more numerous in any locality than are weasels, raccoons, snakes, and foxes. The animals at the base of the food chain are abundant, while those at the end are relatively few in number. There is a progressive decrease in between the two extremes. The small herbivorous animals in the community are able to increase at a very high rate, which provides a large margin of numbers over and above that necessary to maintain their population in the absence of enemies. This surplus supports a set of carnivores, which are larger in size and fewer in numbers. Finally, a point is reached at which a carnivore cannot be supported by any further stage in the food chain. This arrangement of numbers in the community with a relative decrease in population densities at each stage in the food chain is characteristic of animal communities all over the world. It was termed by Charles Elton[6] 'the pyramid of numbers'.

The relative population densities in a food chain follow the Second Law of Thermodynamics. The mass of herbivores cannot be as great as the mass of plants they feed on. With each step upward in a food chain the biomass is reduced. Energy present in the chemical bonds of organisms at one level does not all end up as bond energy at the next level because much of the energy is degraded to heat at each step. The organisms at each level in the chain require energy to grow and maintain life. This energy is not captured by the predators. Predators that capture their prey efficiently without flight or struggle are less wasteful of energy than those that chase their prey.

Pyramid of Numbers

In general, the concept of food chains and food cycles in the pyramid of numbers implies regulation largely by the carnivorous habits of animals in progressively higher stages of the pyramid. If this were the case, studies of communities should indicate stability in numbers, and in biocycles with relatively constant inputs of energy, populations of different species in different parts of the pyramid should fluctuate around a characteristic mean density. This is by no means true, since field studies have proven that the number of animals never remains constant for very long and usually fluctuates considerably from year to year.

Population Census

The measurement of animal populations is a difficult and complex problem requiring large expenditures of labor and finance. The enumeration of the animals of a given area at a given time is called an animal census. The determination of trends or fluctuations requires the census in a given area two or more times, while the determination of relative abundance involves a census in two areas at the same time. The best source of information about population densities comes from the literature on game management, as game animals are a natural resource of considerable value.

Population Curves

A graph illustrating numbers relative to time is termed a population curve. Aldo Leopold, credited with founding the science of game management in America, describes three types of population curves: flat, cyclic, and irruptive (Fig. 6). The flat type is characterized by the absence of severe fluctuations and relative stability in numbers year after year. The second type exhibits pulsations with the same regularity as such natural phenomena as phases of the moon or the ebb and flow of ocean tides. Cyclic population curves are considered characteristic of birds and mammals in the northern latitudes. The third type of population curve exhibits severe but irregular fluctuations of no fixed length or amplitude.

Annual Population Curve

Animals seldom breed continuously throughout the year in any part of the world. Instead the breeding season is confined normally to periods of relatively warm weather and abundant food supplies. In the tropical forests, the breeding season is synchronized with rainfall. The result is an annual population cycle with highs and lows (Fig. 7). The low point in the temperate and arctic zones of the northern hemisphere corresponds to the end of winter. Thus the flat population curve more clearly refers to situations wherein the breeding population is close to the same size each year even though the number of young produced each year varies considerably. Because the population returns to the same point at the end of the winter, numbers are said to 'fit' the

Figure 6. Types of Population Curves — Modified from Leopold[7]

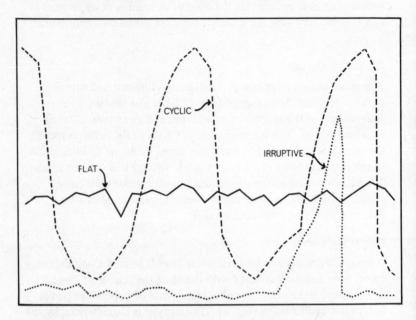

carrying capacity of the range or habitat. The surplus produced, that portion exceeding the carrying capacity of the habitat, always perishes, consumed by predators or disposed of through disease and starvation. Carrying capacity is a favorite concept of game managers because the surplus is considered a standing crop representing the proportion that can be harvested by hunting and fishing.

Irruptive Population Curve

The irruptive population curve is exemplified by occasional plagues or epidemics of insects, rodents, lagomorphs, and in some instances, of ungulates. The occurrence of irruptive phases results from irregularities of the environment which at times combine to produce especially favorable conditions for population growth. It is axiomatic that no population can continue to grow unchecked indefinitely, thus the irruptive stage is always followed by a cataclysmic 'crash'. The causes of these rapid collapses of the populations are classically listed as disease, parasitism, starvation, and 'stress', but more about causality later. The important principle to be assimilated now is the historical significance of the observations of violent fluctuations in the numbers of many species. Such phenomena stimulated

Figure 7. Annual Population Cycle of the Woodchuck *(Marmota monax)* in
Central Pennsylvania. Three Different Study Areas and Three Different
Methods of Estimating Population Size. Area C — 600 Acres, Area D — 535
Acres and Area G — 333 Acres.

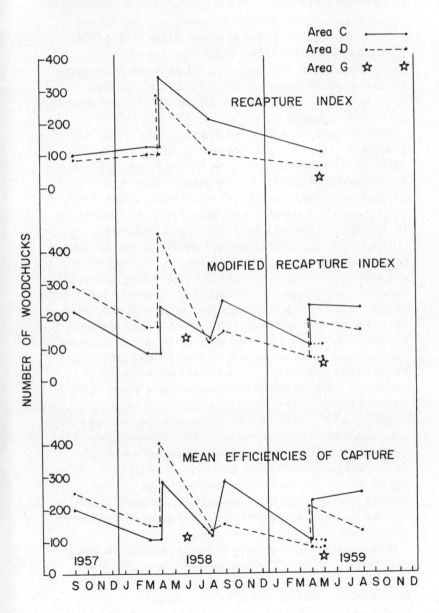

biologists to search for inclusive laws of population which fit the data gleaned by the naturalists.

Cyclic Population Curve

The cyclic type of population curve was even more of a stimulus to research, since periodic oscillations of more or less fixed amplitude suggested adherance to an as yet undiscovered natural law. Arguments about the authenticity of cyclic phenomena have raged for many years and are not yet settled, but again the historical significance of the cycle controversy was the impetus it provided for population research.

It is not entirely clear whether cyclic phenomena were recognized generally by naturalists in different parts of the world during the eighteenth and nineteenth centuries or whether this distinction should be attributed retrospectively to investigators of more recent vintage. Nevertheless, Charles Elton[8] should be given much of the credit because he compiled so much data on the subject and analyzed the material critically and scientifically. Elton considered the Norwegian lemming *(Lemmus lemmus)* and the Canadian varying hare *(Lepus americanus)* as the most striking examples of species whose fluctuations were extraordinarily regular in their rhythm. The lemming especially appears to have attracted attention for hundreds of years because of its periodic appearance in tremendous numbers and its propensity to migrate over even a hundred miles of country into areas not normally considered its abode. Upon reaching the sea, these creatures are not turned back but swim on until they become exhausted and drown. Naturally such preternatural behavior was the basis for myth and legend in the past but, more important, it presented a challenge to the biologist to explain the inexplicable. According to Elton, in recent years the maxima have occurred every four years, while in the middle of the nineteenth century they sometimes occurred at shorter intervals of three or even two years. By examining the records of fox skins obtained by the Hudson's Bay Company in Canada, Elton was also able to ascertain that lemming years in Canada and Norway synchronize almost exactly, a curious fact which he interpreted to mean that lemming numbers were controlled by climate.

After investigating the evidence then available for North America, Aldo Leopold, obviously a believer in cycles, offered a set of nine postulates in his book on game management published in 1948:
(a) Gallinaceous game and rabbits are normally flat in the center or

optimum of their indigenous ranges, but tend to become cyclic or
fluctuating as they approach the geographic limit of their distribution.
(b) When their distribution is artificially extended to acquired range,
their population curves become more cyclic. (c) Cycles are more
severe on large continuous blocks of range than on small, dispersed,
or discontinuous blocks. (d) There is no evidence of zonal distribution
of incidence, except possibly in very large areas. (e) The length of
the cycle period in North America averages about ten years and is
apparently somewhere near synchronous in the various parts of the
continent. (f) The length of the cycle period in the British Isles
averages about 6.5 years and is somewhere near synchronous in its
various parts, but (as is evident from its length) not synchronous
with the North American period. (g) The fur bearers of Canada
in general share the grouse and rabbit cycle, except that the muskrat
cycle is inverse to it. (h) The lemmings and Arctic white foxes
of the American Arctic show a four-year cycle. (i) Salmon and cod
on the Atlantic coast of Canada seem to share the grouse and rabbit
cycle. Leopold[7] concluded that the cause or causes of wildlife cycles
are unknown, but that the nine postulates drawn from their behavior
in time and space at once narrowed the possible field of speculation.
The seeming synchronism of cyclic phenomena led him to
postulate some cause operating simultaneously over the whole
North American continent. Fluctuations in solar radiation, in
electro-magnetic conditions, or in some other cosmic force were
thought possibly to meet this specification.

Aldo Leopold was not alone in advocating the reality of cyclic
phenomena. The Matamek Conference on Biological Cycles in 1932
was devoted exclusively to this problem. Trippensee[9] in 1948
remarked that a current of mysteries still obscures a clear view of how
the rise and fall of animal populations occur and why animals
unrelated in form or physiology are affected. He presented a list of
seven clues which fairly well exemplifies the generally accepted
conclusions up to that time: (1) Cycles differ in length of time,
depending on the size of the animals. (2) Cycles are more pronounced
in the zone of severe cold climates where snow is present part of
the year than where warm seasons without snow are continuous and
milder weather prevails. (3) Herbivores are more subject to cyclic
change than are omnivores. (4) Animals that can control their
environment to a limited degree and consume more than the current
plant growth, such as the beaver, are less subject to cycles than those
which have no control over their environment. (5) Cycles are not the

results of basic changes in the breeding mechanism such as the number of young in a clutch or litter or the number of litters. The crash occurs when a high proportion of the yearly increment is lost, such as the eggs of birds and the young of mammals. (6) 'Carnivores respond to variations in their food supply rather than being the cause of the variation in the number of prey.' (7) The force that controls the various cycles may be a cosmic force, but as yet its relation to the secondary cause or causes of mortality has not been satisfactorily explained.

Sunspots

A search for extrinsic or extramundane cosmic forces led DeLury[10] to suspect that game cycles were caused by sunspot variations and the corresponding changes in solar radiation and weather. Sunspots are local areas of the sun's surface which are darker and cooler than the rest of the surface. Some sunspots are large enough to be seen without a telescope, the larger spots reaching a width of from 60,000 to 90,000 miles. The visible surface of the sun is called the photosphere. The dark inner region of a sunspot called the umbra is about 1,200 degrees lower than the 6,000 degree temperature of the photosphere. The number of sunspots counted in a single year varies between maximums and minimums in a cycle of about eleven years. The number counted in 1947, a year of maximum sunspot activity, was 664. The next minimum year, 1954, produced a count of only 46. An examination of data gathered from astronomers reveals that sunspot periodicity is not reproduced with clock-like precision. A minimum interval of 7.3 years and a maximum interval of 15.0 years are noted. Years of high sunspot activity bring increased heat radiation to the earth, which in turn causes greater evaporation on the surface of the earth, greater precipitation, and cooler conditions. Thus many biologists postulated that cyclic behavior in some species was due to climatic influences brought on by variations in sunspot activity. The sunspot theory was attractive because it fulfilled certain predicated conditions — extramundane or cosmic forces that could account for the population changes over large geographic areas. Ecologists who considered population density to be primarily related to the available food supply were apparently satisfied because plant growth and seed production were directly influenced by climatic conditions.

Reality of Cycles

The reality of cycles was subsequently questioned by some biologists. The highs and lows of the population curves did not always reflect

favorable or unfavorable weather conditions and the sunspot
periodicity seldom fitted with precision. Also, apparently some of the
data analyzed was subjected to smoothing and other manipulations
that produced a superficial appearance of cyclic patterns. Lawrence
Slobodkin[11] illustrated mathematically how cycles can be simulated
by inaccurate censuses. In this case the mean cycle length depends on
the number of population levels recognized (n):

$$\frac{6n^2}{(n-1)(2n-1)} .$$

Palmgren[12] and Cole[13] have pointed out that the peaks of
population fluctuations are random, which does not mean that the
oscillations are not causal, but that the causes vary in a random manner.
Cole[14] states that in most cases 'natural populations are not subject
to alteration only by two or three interacting factors but are affected
by a large enough number to encourage us to look for secondary
simplicity; the highly acclaimed regularity of population cycles seems
to be no greater than that which is encountered in a sequence of
random numbers.'

The general conclusion drawn from mathematical and statistical
analyses of this sort is not that populations lack periodic highs and
lows, but that the highs and lows occur in a random fashion and are
not likely to be related to rhythmic, oscillating cosmic forces.

Although cyclic phenomenon may have attracted more than its
share of attention, the fact remains that animal populations
characteristically exhibit significant growth phases and corresponding
crashes in certain geographic locations which still must be considered
in light of regulatory factors. Accounts of irruptions and drastic
declines are legion. The American mammalogist William Hamilton
reported the variation in numbers of jack rabbits (Lepus sp.) on a
square mile area as a few during a low to 69 during a high. According
to Criddle,[15] the snowshoe rabbit may reach 64 per square mile during
a low, and 3,200 in the same area during a high. Green and Evans[16,17,18]
found a high of 500 hares per square mile in February 1933 and a low
of 32 animals per square mile five years later on a 6.5 square mile
area near Lake Alexander, Minnesota. The number of ruffed grouse
may fluctuate from six to eight per square mile for early spring during
a low to 50 or more for the spring season during a high. Hamilton[19]
found that field mice may vary from a low of one or two dozen per
acre to several hundred per acre during the high of a cycle. During a

mouse outbreak in California, the numbers at the height of the outbreak were estimated at 80,000 per acre, but this probably should be more correctly termed an irruption rather than one of the phases of a cycle.

Factors Influencing Population Growth

Population growth for any species cannot continue indefinitely because prerequisites of the environment would ultimately be exhausted. In recognition of this fact Justus Liebig in 1840 formulated the law of the minimum, which states that the prerequisite in least amount will limit population growth.[20] Fluctuations in population density, at least in vertebrate populations, usually occur at levels well below the limits set by the minimal factor, which suggests that other factors are operative. However, such factors do not truly limit population growth and are termed regulatory.

In a discussion of the questions of mortality in insect populations, Howard and Fiske[21] divided mortality into two categories — catastrophic and facultative. Catastrophic referred to factors that destroyed a constant percentage of the insects irrespective of the level of density and facultative to factors that destroyed a percentage increasing as the number per unit area increased. Smith[22] introduced the terms 'density-independent' and 'density-dependent' to describe mortality factors. Eventually factors that affected reproduction and dispersion were classified in the same manner. The effects of physical factors such as climate, hydrogen ion concentration, pollution, and nutrition may be profound and catastrophic but for the most part independent of density. Intraspecific and interspecific competition, predation, parasitism, and disease are examples of density-dependent factors because they may exert a greater effect when density is high. A.J. Nicholson, an Australian economic entomologist, incorporated density dependency into his theories of population regulation.[23] In discussing the concept of balance in nature, Nicholson decided that it was essential that a controlling factor should act more severely against an average individual when the density of animals is high and less severely when the density is low. In other words, the action of the controlling factor must be governed by the density of the population controlled. Controlling factors in Nicholson's opinion were always competition — competition for food, competition for a place to live, or the competition of predators or parasites.

Andrewartha and Birch,[24] two Australian zoologists, disagreed completely with Nicholson. They rejected the distinction Howard and

Fiske[21] had made between physical and biotic factors and also the classification of the environment based on density-dependent and density-independent agents. They believed that all environmental factors were dependent on population density. According to Andrewartha and Birch, the numbers of animals in a natural population may be limited in three ways: (a) by shortage of material resources, such as food, places in which to make nests, and so on; (b) by inaccessibility of these material resources relative to the animal's capacities for dispersal and searching; and (c) 'by shortage of time when the rate of increase is positive'. On the face of it, the third category seems a bit obscure, but these zoologists contended that any component of the environment (e.g., weather, predators) causes fluctuations in the rate of increase.

The English ornithologist David Lack agreed with Nicholson's viewpoint that 'stability' or 'balance' must be brought about by density-dependent factors.[25] Lack concluded from studies of birds that food shortage was the most important factor regulating population and that it operated through density-dependent changes in mortality operating chiefly on juveniles.

Figure 8 is a schematic representation of the principle of density dependency. Reproduction, mortality, and dispersal can be independent of population density (line A), but under such circumstances the environmental agents affecting them do not control or regulate population growth. An agent affecting reproduction in a density-dependent manner will regulate growth if the rate of recruitment to the population is inversely proportional to population density (line B). Conversely, an agent affecting mortality in a density-dependent manner will regulate population growth if mortality rate is directly proportional to population density (line C). Density-dependent emigration will regulate population growth if proportionately more animals leave the population at higher densities (line C). Density-dependent immigration will regulate population growth if the rate of immigration is higher at low densities (line B). Conceivably there might be circumstances at which reproduction or immigration (recruitment) would be greater at higher densities. If so, self-regulation would not apply and the population would continue to grow until another regulatory factory became operative.

Extrinsic Regulation of Populations

Ironically, an eighteenth-century English country gentleman of

Figure 8. The Principle of Density Dependency

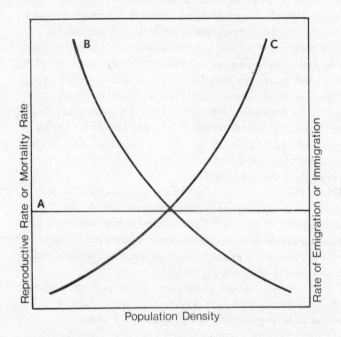

independent means, Thomas Robert Malthus (1766–1834), focused
attention on the problem of food supply and population growth.
His father was well-connected in the intellectual and philosophical
circles of the time, being a close associate of David Hume and
a correspondent and friend of Jean Jacques Rousseau. Malthus
graduated with first-class honors in mathematics from Cambridge
in 1788. With graduation he took Holy Orders in the Church of
England and subsequently earned his MA from Cambridge in 1791
and became a Fellow of the College in 1793.

Malthus' famous first essay was published anonymously in 1798
under the title, 'An Essay on the Principle of Population as it
Affects the Future Improvement of Society With Remarks on the
Speculations of Mr. Godwin M. Condorcet and Other Writers.'
William Godwin had published his 'Enquiry Concerning Political
Justice' in 1793 and the Marquis de Condorcet his 'Essay on the
Progress of the Human Spirit' in 1794. Both gentlemen theorized
that mankind was about to enter a bright new age of enlightenment
and reason. All discord, want, and cruelty were held to result

from an ignorance of immutable Laws of Nature. Strict compliance
with these laws would solve the age-old problems of war, famine, and
misery.

Thomas Malthus did not share the optimism of the intellects of his
day. His first essay centered on the proposition that man's 'power of
population is indefinitely greater than the power in the earth to
produce subsistence.' His mathematical training no doubt influenced
his presentation '. . . population, when unchecked, increases in a
geometrical ratio. Subsistence increases only in an arithmetical
ratio . . .' Reactions to the first essay were predictably hostile. The
controversy led to the publication in 1803 of a second essay, less
speculative and more documented. This one was signed and titled,
'An Essay on the Principle of Population or a View of its Past and
Present Effects on Human Happiness with an Inquiry into Our
Prospects Respecting the Future Removal or Mitigation of the Evils
it Occasions.'

Malthus is credited with influencing Charles Darwin, who based
his theory of evolution on the struggle for existence and competition
between organisms. Darwin and Malthus, more than anybody, were
probably responsible for the emphasis that population students placed
on such extrinsic regulatory factors as subsistence levels, predation,
and interspecific competition during the first half of the twentieth
century. Malthus believed in catastrophic events as ultimate limiting
factors, an idea that is still existent in certain modern theories of
population control.

Intrinsic Regulation of Populations

The most active fields of research since about 1950, and certainly
the most controversial, involve the question of intrinsic regulation of
population density. In their textbook on animal ecology Allee and
co-workers[2] wrote that fluctuations in population levels are obviously
caused both by physical and biotic factors. 'Density-independent
factors appear to set the limits of possible abundance for a population
in its physical habitat, besides inducing particular changes in density;
while density-dependent factors, operating primarily through
competition, appear to regulate population density through time,
thereby achieving whatever condition of equilibrium or"balance"actually
obtains.' Before 1950 the concept of intrinsic regulation was limited
to theoretical discussions.[23, 22] The frequency with which the terms
'crowding' and 'overpopulation' were used would create the
impression that facilitation of the spread of disease or parasites and the

inhibitory effects of the contamination of the environment by excreta were the intrinsic mechanisms visualized. For the most part ecologists in this era did not consider psychological and physiological characteristics of the member organisms of a population as having any role in regulating numbers. In this context it is probably significant that the authors of *Principles of Animal Ecology*[2] commented on the tendency in the 'older literature' to think of crowding in psychological terms and to attribute causation to something unique in the number relationship themselves '. . . this has led at times to a certain mysticism.' They recommended that such an explanation should not be advanced until other reasonable possibilities have been excluded.

The actual mechanisms by which intrinsic regulation can be achieved in natural populations involve some form of mutual interference between individuals or intraspecific hostility in general. Social behavior, which is at least partly genetic, is important to most theories of natural regulation. The tendency of vertebrate animals to defend territories, for example, would limit population density because dominant animals tolerate conspecific neighbors only within certain minimum spacing patterns. The physiological and endocrinological mechanisms controlling reproduction and resistance to pathogenic agents and parasites are affected by external stimuli. This interaction between the internal environment of the organism, 'the internal milieu,' and the external environment is now established unequivocally. Psychological pressures, acting through the central nervous system (hypothalamus) and certain endocrine glands, are capable of affecting reproductive function and immunological mechanisms. 'Intraspecific hostility' or 'social interaction,' greater in dense populations, triggers psychological responses or pressures which can act on reproduction and immunological defense mechanisms. Hence psycho-social interactions are capable of acting on all three parameters of population growth – reproduction, mortality, and movement.

General Theories of Natural Population Regulation

Although theories of natural regulation can be divided generally according to the emphasis placed on extrinsic and intrinsic factors, the arguments are at first sight still confusing. A recapitulation describing the various schools of thought on the subject should be helpful at this point.

Since populations of plants and animals do not increase without limits but show more or less restricted fluctuations, two questions may be raised. What stops population growth and what determines average abundance? The interactions between the population and the environmental factors of weather, food, shelter, and enemies (predators, parasites, pathogenic organisms) were the only relationships considered by the earlier theorists. A biotic school believed that density-dependent factors were critical. Natural enemies were postulated as the chief density-dependent factors in many populations. A climate school emphasized the role of climatic factors affecting population size and suggested that weather acted as a density-dependent control. A comprehensive school thought that all factors were important, both density-dependent and density-independent ones, and that population changes were controlled by a complex of biotic and physical agents varying in location and time.

By contrast, the self-regulation school focuses on the interactions of conspecific members of an animal population, on differences in their behavior and physiology. The general premise of this school is that reproduction, mortality, and dispersal rates change because individuals are affected by population density. Social interaction, intraspecific competition, intraspecific intolerance, and genetically determined behavior patterns are intrinsic sources of stimuli which act through the central nervous system to induce changes in physiological and endocrinological functions. Reproduction and defense mechanisms against disease and parasites, of course, are dependent on physiological and endocrinological mechanisms. Social interaction, intraspecific competition, and intraspecific intolerance are believed to be density-dependent and related to population size in a positive manner (line C in Figure 8). Genetically determined behavior patterns may deteriorate at higher densities (line B in Figure 8).

Another school of intrinsic regulation might best be described as the genetic school. Members of this school believe that average abundance may be altered by genetic changes in populations. The best example of genetic changes regulating population size is the myxomatosis — rabbit interaction in Australia. The high rabbit populations of the 1940s were decimated by the introduced myxomatosis virus during the early 1950s. Now the virus has evolved so that attenuated strains have replaced the virulent strains, and in addition the rabbits are more resistant to this disease. The rabbit population in Australia is now much smaller than it was before the virus was introduced.

The theories of natural regulation of numbers (both extrinsic and intrinsic) are not mutually exclusive but overlap, and a synthesis of several approaches will be attempted in this book. In the following chapters we shall look at the experimental data on which these theories are based and also at the results of studies of natural populations.

Natural Selection

At this point we should consider the possibility that individuals acting in concert and co-ordinated groups of animals might be subject to natural selection as units. Intraspecific population units at various levels of integration are recognized and referred to under various terms: the species composed of individual organisms, the aggregation, school, flock, herd, the sex pair with sexual dimorphism, the family with reciprocal adjustments between parents and offspring, and the society in the strict sense with marked division of labor between adults of the same sex.[2] The existence of complex internal adaptation between parts of an organism and integrated homeostatic systems (nervous system, endocrine system, enzyme systems, etc.) within an organism is explicable only through the action of selection upon whole units. Conversely, these integrated groups of individuals would not exist as entities unless natural selection acted upon each whole system.

Consequently the natural regulation of population size as an adaptive organization of the group (i.e. evolution of mechanisms to regulate density) has also been proposed.[26, 27, 28] The reasoning is straightforward. Uncontrolled population growth would tend to diminish the food supply produced by plants to a critical level. Groups of animals within certain restricted geographical ranges would be eliminated by starvation in times of crisis, while those in adjacent ranges that have limited their numbers by biotic negative feedback mechanisms would survive. Immigration of animals from the surviving populations or groups into the ranges thus vacated would provide for the survival of individuals with physiological and psychological mechanisms that respond to intraspecific interactions. Critics of this type of evolutionary thought believe that the selection of alternative alleles in mammalian populations can produce adaptive organization of individuals in family groups. Adaptive organization of a population must be attributed to the selection of alternative populations. Organic adaptations, which function to maximize survival of individuals, are distinguished from biotic adaptations, which perpetuate a population or more inclusive group.

Summary and Conclusions

The first chapter in any book should ultimately introduce the reasons for writing the succeeding chapters. Therefore, why be concerned about the place of population in the science of ecology? The population problem would be of interest to only a handful of people except that the human species is also part of a population. We should consider whether the biological principles applied to populations of animals can teach us anything about our own population problem.

There is an understandable tendency in any synthesizing discussion to concentrate on the progress made in recent years or decades. But in doing so, excellent work of previous decades or even centuries may often be neglected. A false idea of rapidity of progress is encouraged and the supposition that the subject matter is entirely modern is built up in the mind of the reader. Malthus is no doubt responsible for identifying the population problem for humans, but his ideas were not entirely new. Machiavelli, 275 years before Malthus, had realized the danger that human populations may increase beyond the means of subsistence in limited areas, and that such an increase would then be checked by want and disease. Many other writers between 1590 and 1770 anticipated Malthus. In fact, in a book called *The Primitive Origination of Mankind,* published in 1677, the author[29] stated that the increase in the human population tends to occur in geometrical ratio.

In the succeeding chapters I will attempt to assimilate the results of the historical investigations of population growth with more recent research. In particular, the effects of the physical environment and biotic factors on the forces of population growth (reproduction, mortality, and dispersion) will be considered in more detail. Wherever possible I will use data collected from my own field studies to illustrate the discussions. As I mentioned previously, there is much controversy surrounding the population problem, and there is little unanimity of opinion with respect to such important propositions as the existence of natural regulation of population density and whether selective forces are in any way involved with adaptive mechanisms. My investigations of population dynamics, using both experimental populations in the laboratory and natural populations, have involved a search for the ways in which members of populations compensate for changes in density and population structure. The last chapter in the book will consider the implications of past and current population research on the human population problem.

Notes

1. W.C. Alee and K.P. Schmidt, 1951, *Ecological Animal Geography.* 2nd ed., John Wiley and Sons, Inc., New York, Chapman & Hall, Limited, London.
2. W.C. Allee, A.E. Emerson, O. Park, T. Park and K.P. Schmidt, 1949. *Principles of Animal Ecology.* W.B. Saunders Company, Philadelphia and London.
3. G.M. Edelman, 1970. 'The Structure and Function of Antibodies'. *Scientific American,* Aug., pp.34-53.
4. E. Mayr, 1942. *Systematics and the Origin of Species.* Columbia Univ. Press, New York.
5. K.M. Taylor, D.A. Hungerford, R.L. Snyder and F.A. Ulmer, 1968. 'Uniformity of karyotypes in the Camelidae', *Cytogenetics,* 7:pp.8-15.
6. C. Elton, 1927. *Animal Ecology.* MacMillan Co., New York, Sidgwick and Jackson, London.
7. A. Leopold, 1948. *Game Management.* Charles Scribner's Sons, New York, London.
8. C. Elton, 1924. 'Periodic fluctuations in the numbers of animals: their causes and effects'. *J. Exper. Biol.* 2 : pp.119-63.
9. R.E. Trippensee, 1948. *Wildlife Management.* McGraw-Hill Book Company, Inc., New York, Toronto, London.
10. R.E. DeLury, 1930. *Sunspots and Living Things.* Trans. 17th American Game Conference, pp.211-12.
11. L.B. Slobodkin, 1961. *Growth and Regulation of Animal Populations.* Holt, Rinehart, and Winston, New York.
12. P. Palmgren, 1949. 'Some remarks on the short-term fluctuations in the numbers of northern birds and mammals'. *Oikos,* 1 : pp.114-21.
13. L.C. Cole, 1951. 'Population cycles and random oscillations'. *J. Wildl. Mgt.* 15 : pp.233-51.
14. ——— 1957. 'Sketches of general and comparative demography'. *Cold Spring Harbor Sym. Quant. Biol.,* 22 : pp.1-15.
15. S. Criddle, 1938. 'A study of the snowshoe rabbit'. *Canad. Field Nat.* 55 : pp.31-40.
16. R.G. Green and C.A. Evans, 1940. 'Studies of a population cycle of snowshoe hares on the Lake Alexander area. I. Gross annual censuses, 1932–1939'. *J. Wildl. Mgt.* 4 : pp.220-38.
17. ———1940. 'Studies of a population cycle of snowshoe hares on the Lake Alexander area. II. Mortality according to age groups and seasons'. *J. Wildl. Mgt.* 4 : pp.267-78.
18. ———1940. 'Studies of a population cycle of snowshoe hares on the Lake Alexander area. III. Effect of reproduction and mortality of young hares on the cycle'. *J. Wildl. Mgt.* 4 : pp.347-58.
19. W.J. Hamilton, Jr.,1939. *American Mammals – their Lives, Habits, and Economic Relations.* McGraw-Hill Book Company, Inc., New York.
20. J. Liebig, 1840. *Chemistry in Its Application to Agriculture and Physiology* (4th ed. in 1847). Taylor and Walton, London.
21. L.O. Howard and W.F. Fiske, 1911. 'The importation into the United States of the parasites of the gipsy moth and the browntail moth'. *Bull. U.S. Bur. Entom.* 91 : pp.1-344.
22. H.S. Smith, 1935. 'The role of biotic factors in the determination of population densities'. *J. Econ. Entom.* 28 : pp.873-98.
23. A.J. Nicholson, 1933. 'The balance of animal populations'. *J. Adim. Ecol.* 2 : pp.132-78.
24. H.G. Andrewartha and L.C. Birch, 1954. *The Distribution and Abundance of Animals.* University of Chicago Press, Chicago.

25. D. Lack, 1954.'The Natural Regulation of Animal Numbers'. Oxford University Press, New York.
26. J. Le Gay Brereton, 1962. 'Evolved regulatory mechanisms of population control'. In *The Evolution of Living Organisms,* G.W. Leeper, ed., Melbourne University Press.
27. R.L. Snyder, 1961. 'Evolution and integration of mechanisms that regulate population growth'. *Proc. Nat. Acad. Sci. U.S.* 47 : pp.449-55.
28. V.C. Wynne-Edwards, 1962. *Animal Dispersion in Relation to Social Behaviour.* Oliver and Boyd, Edinburgh and London.
29. M. Hale, 1677. *The Primitive Origination of Mankind.* Shrowsbery, London.

2 POPULATION GROWTH – REPRODUCTION

Reproduction, the process by which living organisms give rise to offspring, is also the positive force of population growth. Mortality is always a negative force, the population-decline factor, and thus antithetic to reproduction or natality. Movement can augment growth if immigrants settle within the geographical limits of the population.

Biologists have long recognized that living organisms possess a tremendous potential for increase, which if unchecked would soon inundate the world with protoplasm. For a number of reasons, which will be discussed in this chapter, this potential reproductive rate is seldom realized. On a population basis it is fair to say that the potential rate is never realized.

Part I: Types of Reproduction

Asexual Reproduction

It is useful to describe briefly the types of reproduction existing in various species of plants and animals and to be somewhat familiar with the physiological and endocrinological mechanisms involved in vertebrate reproduction as a prerequisite for understanding how various extrinsic and intrinsic factors can influence population growth. Reproduction that involves only one parent and no special reproductive structures is termed asexual reproduction. This simplest of types occurs in many plants and fungi and in protozoans, coelenterates, bryozoans, and tunicates. Bacteria and protozoa usually multiply by binary fission in which an individual divides into two halves, usually equal, after which each grows to the original form. A type of multiple fission, or sporulation, involves repeated divisions of nuclear material with subsequent apportionment of the cytoplasm so that it surrounds each of the new daughter nuclei. Budding is a type of reproduction in which a new individual arises as an outgrowth on an older animal. In some flatworms and ribbon worms an individual may fragment itself into two or more parts, each capable of growing to be a complete animal.

Sexual Reproduction

Most plants, fungi, and many kinds of animals increase by sexual reproduction in which new individuals develop from specialized cells produced by the parents. Typically two sex cells of different kind (male and female) join to produce a new individual. Among protozoans there are several reproductive processes closely akin to sexual reproduction in higher animals. In the conjugation of ciliates *(Paramecium,* etc.) two individuals fuse together, exchange micronuclear material, and then separate to continue binary fission. Among the sporozoans *(Plasmodium* of malaria, etc.) two kinds of individuals, macrogametes and microgametes, are produced which fuse permanently in pairs to continue the life cycle.

In multicellular animals, sex is considered the total of all structural and functional characteristics that distinguish males and females. The male produces minute free sex cells, or germ cells, known as sperms or spermatozoa. The female releases somewhat larger sex cells, the eggs or ova. Besides the necessary differences in reproductive organs (gonads), individuals of the two sexes differ otherwise in external and internal form, in physiology, and in behavior.

Parthenogenesis

Development of an egg without the entrance of a sperm is known as parthenogenesis, in which case the succeeding generations consist solely of females. This form of reproduction is advantageous in allowing organisms to exploit brief periods of favorable conditions. The population can be expanded rapidly before inimical conditions return. It occurs in rotifers, plant lice, many ants, bees, wasps, some crustaceans, and certain other invertebrates. Parthenogenesis suffers from the grave disadvantage of genetic invariability, since all the progeny of each haploid ancestral individual are genotypically identical. Usually sexual reproduction intervenes at least once in every annual cycle, allowing genetic recombinations to take place.

Paedogenesis

The larvae of the gall fly *(Miastor* spp.) produce eggs that develop parthenogenetically to yield other larvae. In the liver fluke one larval stage, the sporocyst, produces unfertilized eggs that give rise to another larval stage, the redia. Such instances of parthenogenesis are termed paedogenesis.

Hermaphroditism

If both male and female reproductive systems occur in one individual, the animal is said to be monoecious. The term applied when each individual is either male or female is dioecious. The term hermaphrodite is applied to monoecious species and also to the occasional abnormal individuals of dioecious species that contain both male and female reproductive systems. All adult hermaphrodites are oviferous (i.e. capable of producing eggs), but cross-fertilization between individuals can of course proceed in the normal manner. A condition known as protandry wherein one gonad alternately produces eggs and spermatozoa occurs in some oysters and the hag fishes. Most hermaphrodites produce ova and spermatozoa simultaneously, as in the earthworms and most of the common slugs and snails. In the absence of cross-fertilization, some hermaphrodites can be self-impregnating. A closely analogous form of hermaphroditism is by far the commonest condition found through the whole range of higher plants.

Evolution of Reproductive Mechanisms

During the course of evolution of multicellular animals, there have been marked anatomical, hormonal, and physiological changes, which are assumed to have enhanced the protection of the young and the survival of the species. In general these changes involved economy in the production of gametes, reduction in the size of the egg, internal fertilization, development of the corpus luteum as a temporary endocrine organ, and development of the placenta as a nutritive, excretory, endocrine, and protective organ. Thus, placental mammals might be considered to have reached the pinnacle, as far as efficiency of reproduction and adaptability are concerned. Presumably each innovation in reproduction and care of the young resulted in greater survival – recruitment to the population.

When considering reproduction as a population attribute, one is required to understand the reproductive processes of the species in question and how environmental factors affect various stages of reproduction. Each species must be considered separately, as no two species have exactly the same physiological and endocrinological pathways or the same reproductive behavior. Our concept of the niche includes the belief that all of these differences are individual adaptations which allow the species to fit the niche.

Reproductive Physiology

Reproduction must be integrated with external events, otherwise the species would perish. In most animals, breeding is confined to one specific period of the year, to a season most propitious for the successful birth or hatching of the young and for their subsequent care. Physical environmental factors appear to be of primary importance in determining the time of onset of breeding in most species, although innate cyclic phenomena that condition the organism to respond to external stimuli can also be demonstrated. Seasonal variation in light, for example, is a stimulus for attainment of reproductive activity, especially among birds and mammals. The more familiar domestic birds and mammals carry on reproduction more or less successfully throughout the year. But in this case, selective breeding and protection from the environment have negated the value of seasonal breeding. Even so, expanding the length of daylight by artificial lighting to increase egg production in poultry is a well-accepted practice. The higher primates, including man, generally show continuous breeding.

Maturation of the sexual organs, reproductive cycles, ovulation, spermatogenesis, pregnancy, parturition, and the function of the mammary glands are controlled and integrated by the nervous and endocrine systems. Since the nervous system operates by highly localized release of a chemical, such as acetylcholine, noradrenaline, or serotonin, it is involved in high-speed responses of short duration. The endocrine system, in contrast, is involved in slower, longer lasting responses through secretion of its active chemical, the hormone, into the bloodstream where effective concentrations are maintained for appreciable lengths of time. Both systems depend on delivery of an activating chemical to an effector organ. The systems differ primarily in the way in which the chemicals are delivered. Presently the organismal mechanisms of reproduction are fairly well understood. Exactly how internal physiology is affected by external environmental factors is not.

Endocrine Organs in the Invertebrates

The main endocrine organs of the vertebrates have been recognized for a great many years, and extensive studies of their anatomical and physiological characteristics have been conducted. Our knowledge of endocrine functions in the invertebrates, however, has been for the most part incidental to these studies. Typical experiments involved the study of the effects of vertebrate hormones on various invertebrate species. While this research did little to elucidate natural endocrine mechanisms in the lower animals, it established the fact that invertebrates will respond to hormonal influences.

Secondary sexual characteristics in annelids and molluscs are dependent on endocrine function of the gonads or a male sex hormone originating in another part of the reproductive tract. This phenomenon,

the dependence of the reproductive tract and secondary sexual
characteristics upon gonadal hormones, so well-established in
vertebrates, has not been established in insects. On the other hand,
relationship between the insect's corpus allatum, a glandular organ
located in the head or anterior part of the thorax, and ovarian function
is well established. The adult corpus allatum secretes an egg-ripening
factor. Release of this factor is inhibited by nervous control. In turn,
these inhibitory nerve fibers are influenced by a humoral factor released
by ripe eggs in the uterus. The output of egg-ripening hormone by the
corpus allatum therefore is controlled by a neuroendocrine feedback.
The activity of the corpus allatum is controlled by the brain. The
latter in turn receives its cues from the internal and external milieu
in which the insect finds itself. This is a second-order neuroendocrine
control mechanism in which the corpus allatum represents the
intermediary between central nervous system and effector organs.
Several other peculiar organs of the insects are suspected of being
endocrine in function because of their histological appearance.

In crustaceans, no endocrine organ comparable to the pituitary
or corpus allatum appears to participate in retarding gonadal function.
The gonad-inhibiting hormone originates in neurosecretory cells
within the evestalk and is stored in the sinus gland.[1] In *Octopus* the
optic gland, a structure that adheres to the optic stalk, apparently
fulfills the same function as the anterior pituitary in vertebrates
and the corpus allatum in insects. Sexual maturation in annelids
of the families Nereidae and Nephthyidae is delayed by an
inhibitory hormone originating in neurosecretory cells of the head.[1]

Nervous and Endocrine Control of Reproduction in the Vertebrates

Research in endocrinology along comparative or evolutionary lines has
been in reverse order. That is, elucidation of endocrine mechanisms was
accomplished first with birds and mammals, then attention was turned
to a search for similar or analogous systems in lower animals. It follows
then that more is known about nervous and endocrine control of
reproduction in the higher vertebrates. For this reason it seems logical
to discuss mechanisms among the mammals first and then to speculate
about their evolution in the other vertebrate classes.

The endocrine organs directly concerned with mammalian
reproduction and lactation (uniquely a mammalian attribute) are the
pituitary, the ovary, the testes, the placenta, and in some species, the
uterus. The thyroid glands which control metabolic rate and oxygen
consumption by tissues; the parathyroid glands which condition the

metabolism of calcium and phosphorous; the pancreatic islets which are concerned with carbohydrate metabolism and related functions; and the adrenal glands with a multitude of functions are not directly concerned with reproductive functions but play a secondary role because they control vital physiological functions. For example, absence of the parathyroid glands would interfere with the formation of the egg shell in avian species, although this would be of little consequence since birds so deprived would soon become tetanic and die in convulsions anyway. Steroid hormones from the adrenal cortex can affect certain reproductive functions directly, but such actions are not considered within the mainstream of normal reproductive physiology.

The Pituitary

The mammalian pituitary gland is a relatively small unpaired organ attached by a slender stalk to the floor of the brain. It secretes at least nine hormones, all being protein or peptide in nature — six from an anterior lobe, at least one from an intermediate lobe, and two from a posterior lobe. Three of the anterior lobe hormones affect the ovaries, testes, and mammary glands directly, hence they are called gonadotropic hormones. Prolactin, also called lactogenic hormone and luteotropin, induces proliferation of the mammary glands, initiates milk secretion, and prolongs the functional life of the corpus luteum of the ovary. Luteinizing hormone (LH), also called interstitial cell-stimulating hormone (ICSH), influences the formation of corpora lutea in the ovaries and stimulates the interstitial cells of Leydig in the testes. Follicle-stimulating hormone (FSH) governs the growth of ovarian follicles and functions with LH to cause estrogen secretion and ovulation. The last-named hormone also apparently acts on the seminiferous tubules of the testes to promote spermatogenesis. Oxytocin from the posterior lobe causes ejection of milk in the postpartum mammary gland and promotes contraction of uterine muscle. It may also act in parturition and in sperm transport in the female tract.

Apparently the testes and ovaries have, at best, only limited powers of autonomous regulation, thus their functions depend upon adequate levels of pituitary tropic hormones circulating in the blood. Estrogen produced by the ovaries and testosterone, a strong androgenic or male hormone produced by the testes, inhibit the further release of gonadotropins from the anterior pituitary. This 'feedback' mechanism is important to remember because similar mechanisms are apparently involved in interactions between population pressures and reproduction.

They will be discussed in greater detail later.

The mammalian pituitary *(hypophysis)* consists of an *adenohypophysis* and a *neurohypophysis* which are distinctly different in embryonic origin. The adenohypophysis includes the anterior lobe *(pars distalis)* with its *pars tuberalis*, and the intermediate lobe *(pars intermedia)* (see Fig. 9). The neurohypophysis includes the posterior lobe and

Figure 9. Schematic Drawing of the Mammalian Pituitary

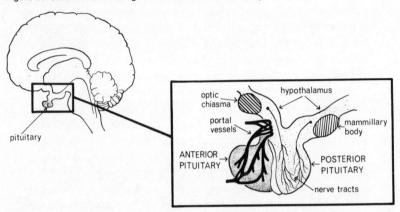

technically also certain nuclei in the hypothalamus and their axons which terminate in the posterior lobe. The adenohypophysis is derived embryologically from Rathke's pouch, an outgrowth from the roof of the mouth, while the neurohypophysis is strictly nervous tissue, having originated from the infundibulum of the brain, an outpocketing of the hypothalamus.

Little is known of the structure or function of possible endocrine organs in the lower chordates. The lamprey, one of the most primitive vertebrates belong to the Cyclostomata, already has a fully developed pituitary. Speculation concerning the evolution of the so-called master gland of the higher vertebrates is interesting although based on tenuous and indirect reasoning. Nevertheless, to appreciate how different stages of reproduction can be affected by external environmental stimuli, the connection between the gonads producing the gametes and the central nervous system must be understood. Neurohemal glands are found among the living protochordate groups as well as in the crustacea, insects, and annelids. These organs consist of endings of neurosecretory cells from the brain gathered together near blood channels. Examples are the sinus gland of the crustaceans and the corpus cardiacum of insects.

In some invertebrates, a glandular structure derived from the nearby epidermis or digestive tract epithelium becomes closely associated with the neural structure.

In their textbook of comparative endocrinology[2] Aubrey Gorbman and Howard Bern suggest a manner in which the mucosal layer adjacent to the neurohypophysis may have evolved into the adenohypophysis. If at some time a mechanism of metabolic responsiveness to neurosecretion developed in the mucous glandular tissue next to the infundibulum, there would be adaptive advantage in enlarging the new 'gland'. The first stages in separation of such a gland from the mucosa might have been the isolation of secretory acini or islets as in the pancreas (actually seen in the hagfish, a cyclostome). Further evolution of the adenohypophysis would involve: (a) consolidation of the gland; (b) differentiation of additional functions by further modification of the hormone; and (c) differentiation of more precise neuroendocrine control by the hypothalamus over the adenohypophysis. In this way evolution would have fashioned a way of integrating environmental events with certain internal physiologic changes. 'By enslaving the adenohypophysis the neurosecretory nervous system has immediately enlarged the number of events it can control, without itself having to manufacture a variety of different neurohumors (hormone) to accomplish the same thing.'

The anatomical features of the pituitary are fairly constant among the higher vertebrates. Perhaps the most significant phylogenetic change was the development of a system of portal blood vessels passing from the median eminence of the brain to the adenohypophysis. The most conspicuous phylogenetic change involved the loss of the pars intermedia in birds and in a few mammals, notably the elephant. The hypophyseal portal system (median eminence to the adenohypophysis) is found in all reptiles, birds, and mammals.

Hypothalamic Control of the Pituitary

The hypothalamus, an area of the diencephalon, lies at the base of the brain, ventral to the thalamus, and forms the floor and part of the lateral walls of the third ventricle. It is bounded anteriorly by the optic chiasma and posteriorly by the mammillary bodies. The median eminence of the tuber cinereum, a specialized expansion of the third ventricle, is connected by a stalk to the pituitary. The anterior pituitary has no appreciable nerve supply,[3] thus direct nervous stimulation of this part of the pituitary to elicit hormone release has always been considered unlikely. The hypophyseal portal system of blood vessels between the median eminence and the pituitary instead has been postulated as the communicating link from the brain. It would appear that hypothalamic nerve fibers secrete substances from their endings into capillaries in the median eminence, which are then carried by the portal system to the pituitary where they stimulate or inhibit the release of the various anterior lobe hormones. This neurohumoral concept of the regulation of secretion of anterior pituitary hormones was patterned after the neurosecretory theory. This theory[4,5] holds that oxytocin and vasopressin are synthesized in the neurosecretory cells of the hypothalamus and transported down their axons to nerve endings in the posterior lobe of the pituitary.

The existence of at least nine hypothalamic regulators (simple polypeptides) of the pituitary is now reasonably well established[3]. Some of these substances satisfy the classical definition of a hormone, but are termed interchangeably releasing factors (RF), releasing hormones (RH), and regulating hormones (RH). They apparently affect the synthesis as well as the release of respective anterior pituitary hormones.

For at least three pituitary hormones there is a dual control system, a releasing factor and an inhibiting factor for the same hormone. The need for hypothalamic inhibitors as well as stimulators of growth hormone, prolactin, and melanocyte-stimulating hormone, can be explained by the absence of negative feedback products from their target tissues. In the case of corticotropin (ACTH), thyrotropin (TSH), LH, and FSH, hormones (corticosteroids, thyroxine, and sex steroids) from the target glands inhibit secretion of these tropic hormones by negative feedback action exerted on the pituitary, hypothalamus, or both.

One hypothalamic hormone, a decapeptide with ten amino acids in sequence, is probably responsible for stimulating the release of both FSH and LH from the anterior lobe of the pituitary.[3] Microgram doses of synthetic LH-RH/FSH-RH cause a discharge of FSH, and LH, in sheep, monkeys, humans, and other species.

The effects of steroids on responses to LH-RH/FSH-RH are extremely complex. Since massive doses of 12 commonly used oral contraceptives containing combinations of estrogen and progestin do not block the stimulatory effects of purified LH-RH on the release of LH in ovariectomized rats, it would appear that the feedback action of contraceptive steroids is exerted chiefly on the hypothalamus or another CNS center rather than on the pituitary.[3] There is some evidence, however, that progesterone and testosterone propionate have some direct effects on the pituitary gland. In general, the thought is that endogenous concentrations of sex hormones can influence sensitivities of the pituitary to gonadotropin releasing factors.

Under experimental conditions, the concentrations of LH and FSH in the plasma can be altered by varying the duration of pituitary stimulation with LH-RH/FSH-RH. When ovariectomized rats previously treated with estrogen and progesterone are given picogram and nanogram doses of LH-RH by rapid injection, the concentration of LH but not FSH in the plasma increases.[6] Intravenous infusion of LH-RH for three to four hours induces a significant increase in the concentration of FSH in the plasma as well.[3] The results of these experiments would explain how a single releasing factor controls the release of two separate gonadotropins. In addition, experiments have demonstrated that natural and synthetic LH-RH FSH-RH can stimulate the synthesis as well as the release of LH and FSH.[7]

The exact mechanism of action of LH-RH/FSH-RH in inducing LH and FSH release is unknown. Adenosine 3^1, 5^1-monophosphate (cyclic AMP) and its derivatives can stimulate LH and FSH *in vitro*.[8] AMP may also be responsible for the action of the polypeptide that induces the release of thyroid stimulating hormone (TSH) from the anterior pituitary.[9]

Control of prolactin release is also centered in the hypothalamus, but there are two hormones involved, prolactin-releasing hormone (PRH or PRF) and prolactin release-inhibiting hormone (PRIH or PIF).[10] The inhibitory influence may predominate in man, rat, rabbit and other mammals and possibly in reptiles and amphibians. In birds the stimulatory influence appears to be of primary importance.[11,12] The chemical nature of these two hormones is unknown, but PRIH and PRH both have low molecular weights.

The Ovary

The mammalian ovary functions chiefly in the production of ova and the elaboration of hormones that regulate the reproductive tract and secondary sexual characteristics and condition mating reactions. Gametogenesis and hormone secretion fluctuate rhythmically during the reproductive span of life. Periodic changes in the female tract determined by the pituitary and ovaries are called *estrous cycles* in subprimate species and *menstrual cycles* in man and other primates.

In most vertebrates sexual receptivity is limited to recurring periods called *estrus*. During heat, or estrus, the female is physiologically and psychologically conditioned to receive the male. *Monoestrous* animals complete a single estrous cycle annually, while polyestrous species complete two or more cycles each year if not interrupted by pregnancy. Ovulation is the dominant event in the estrous cycle and usually occurs during estrus or shortly after it terminates.

Ovulation depends upon the release of gonadotropins from the anterior pituitary. The ovulation-inducing hormone is principally LH, but the exact manner in which it produces rupture of the mature Graafian follicle is still not completely understood. In general, FSH and LH are both required for ovulation, although FSH may accomplish its role mainly by stimulating maturation of follicles to the stage where LH can effect ovulation. Both reflex ovulators such as rabbits which ovulate after copulation and spontaneous or cyclic ovulators such as rats and monkeys are dependent upon a massive discharge of LH signaled by the brain.[13] As mentioned previously, overwhelming evidence supports the concept that this neural impulse is mediated by LH-RH/FSH-RH.

The ruptured follicles after ovulation sometimes become filled with blood to form the *corpus hemorrhagicum.* The granulosa cells accumulate a yellow pigment and transform into *lutein* cells; some theca interna cells, differentiated from connective tissue, also become pigmented and, as *paralutein* cells, contribute to the histogenesis of a new organ, the yellow body or *corpus luteum.* If fertilization occurs, this structure may last for all or part of the period of pregnancy, depending on the species. Whether associated with pregnancy or not, it eventually disappears, and a small fibrous scar, the *corpus albicans* or white body, takes its place.

The ovarian cycle consists of two phases, follicular and luteal. The follicles are the primary source of estrogen, the powerful female sex hormone, and the corpora lutea the primary source of progestogen. The ovary also secretes androgen. The synergistic role of this steroid in the normal female mammal in such events as uterine growth or mammary development is still uncertain. Also, its potential effects on pituitary feedback regulation of ovarian function have not been defined.

The primordial follicles apparently develop without need of pituitary hormones, but maturation of the follicles requires both FSH and LH acting synergistically. LH is not only important in follicular and luteal development, but it also stimulates interstitial cell growth and function in both ovary and testis.

It should be remembered that estrogenic titers in the body fluids are rising during the follicular phase, and that estrogens are acting back upon the anterior pituitary mainly through the hypothalamus or other CNS center. The feedback is not simple inhibition, however.

The administration of small doses of estrogen augments the response of LH-RH/FSH-RH in female rats (not in males).[10,14] This is in contrast to the actions of progesterone and testosterone which have been shown to be inhibitory.[10] Estrogen also stimulates both the germinal epithelium and the corpus luteum in the ovary.

In some mammals the corpus luteum is maintained and induced to secrete luteoid hormone by the third gonadotropin, prolactin, and during pregnancy it can also be stimulated by a luteotropic agent from the placenta.

In birds androgen, as well as estrogen, is able to stimulate ovarian function and it does not appear to inhibit pituitary gonadotropin secretion, according to one authority on the subject. Progesterone stimulates secretion of LH by the pituitary in hens. All of this indicates that the negative feedback action so typical of mammals may be essentially absent from birds.[2]

Attempts to induce ovulation in amphibians with mammalian gonadotropic preparations are often ineffective, while implantation of amphibian pituitary tissues induces ovulation readily. Numerous other agents, such as mammalian growth hormone, LH, FSH, prolactin, and steroids, have been employed experimentally *in vivo* and *in vitro* with conflicting results. In general, there is evidence that gonadotropins act to induce ovulation in amphibians through regulation of ovarian steroid secretion, and that even a pituitary-interrenal axis may be involved. Early stages of egg maturation in fish apparently occur independently of pituitary control. Gonadotropins exert their influence with the beginning of yolk formation.

Full differentiation of the endometrium requires both estrogen and progesterone. Implantation of the fertilized ovum and subsequent placentation are dependent on the presence of the ovaries. As with the pituitary, the ovary is more essential for the maintenance of pregnancy in some species than in others. The ovaries are indispensable at practically all stages for the maintenance of pregnancy in the opossum, mouse, rat, rabbit, golden hamster, 13-lined ground squirrel, goat, and viviparous snake.[15] Ovariectomy during the early months of pregnancy in the monkey and mare does not usually cause abortion. Ovaries need not be present during the second half of gestation in the guinea pig, cat, dog, and ewe.

Failure of implantation and abortion can be prevented in castrated animals by the administration of progesterone. The exact role of estrogen in pregnancy is uncertain, but high levels of the metabolic products of this hormone in the urine of pregnant women and mares suggest that estrogens of pregnancy perform some role peculiar to gestation.[16]

The Testis

The primitive function of the testis was simply the production of male gametes, the spermatozoa. Protochordates have no sex accessory glands and sex hormones secreted by the gonads are unknown. In the vertebrate testis steroid hormones are produced by modified connective tissue cells located between the seminiferous tubules. The predominant testicular steroids are androgenic (masculinizing) substances which are primarily responsible for the maintenance of the accessory system of ducts and glands in the male genital tract and of male secondary sex characteristics. Androgenic hormones are also required for full manifestation of male behavioral patterns.

Pituitary control of the testis is accomplished with two gonadotropins, LH and FSH. LH controls the development and functioning of the interstitial cells. FSH stimulates spermatogenesis and is without effect on the interstitial tissue. The third gonadotropin from the pituitary, prolactin, appears to have no significant influence on the testis.

The Placenta and the Uterus

The placenta and, in certain species, the endometrium, produce gonad-stimulating hormones similar in some respects to those secreted by the anterior pituitary. The placental gonadotropins from different mammals are chemically dissimilar and also have different physiological properties. Chorionic gonadotropin, a glycoprotein, is secreted by the chorionic villi of the placenta of the primate species. Human chorionic gonadotropin (HCG) resembles LH in most of its actions. It converts the corpus luteum of the menstrual cycle into the corpus luteum of pregnancy, thereby prolonging the luteal production of progestogens until the placenta becomes capable of secreting large amounts of the hormones required for the continuation of pregnancy.

The primitive function of the placenta was probably luteotropic, prolonging the life of the postovulatory corpus luteum and thus extending the period of gestation. Further placental adaptation added increased endocrine functions which include the ability to substitute for the ovary. Thus, the placenta also secretes both estrogen and progesterone. In some species, relaxin, apparently a water-soluble protein or polypeptide which causes relaxation of the pelvic ligaments to assist parturition, is secreted by both the ovary and the placenta.

One aspect of placental function that is of particular interest is its apparent autonomy. There seems to be no feedback relationship. The secretion of gonadotropins is not under hypothalamic control, and there is no evidence for the existence of releasing hormones regulating placental endocrine activity.

Part II: Factors Affecting Reproduction

The magnitude of reproduction for a given species varies considerably from season to season in the same geographical locality and also from one geographical locality to another in the same season. This fact has stimulated a tremendous volume of research concerning the factors affecting reproduction. A plethora of research studies has accumulated a long list of external environmental agents that intervene to affect reproductive activities and ultimately adjust recruitment to the population. External physical stimuli, such as light intensity, temperature, humidity, hydrogen ion concentration, oxygen concentration, and pressure have significant effects on reproduction, in some instances limiting recruitment and at times preventing reproduction entirely. It should be remembered that we are concerned in this chapter with only those factors that affect reproduction. Naturally recruitment to the breeding population is diminished if a portion of the breeding stock dies, but such losses are properly tabulated on the negative side of the ledger. Organisms may survive within certain ranges of temperature or oxygen concentration, for example, but they don't reproduce. Reproduction is the positive force for population growth.

On the other side of the coin are the biotic factors that affect reproduction. Biotic factors can be divided conveniently into two categories: (a) those arising from interspecific relationships of organisms; and (b) those arising from intraspecific relationships. It is also convenient to discuss effects in terms of density-dependent and density-independent agents. In general, environmental forces act through the intermediation of the nervous system and whatever endocrine systems have evolved. In organisms without a central nervous system and in those with only primitive endocrine functions, reproduction is possibly at the whim and mercy of the physical environment. In the higher vertebrates environmental forces act through the intermediation of the hypothalamus and the anterior pituitary, the latter functioning as the liaison organ between the nervous system and the target glands. The main gap in our knowledge is how the central nervous system perceives environmental stimuli.

Environmental factors may exert their effects at various stages of the reproductive process and affect either male or female fertility. The following discussion is organized to illustrate generally whether the environmental agents are physical or biotic and dependent or independent of population density. The stages of reproduction affected and the mechanisms involved (neural, hormonal, or physiological) are the important considerations.

Research conducted to explore the effects of environmental forces on reproduction has utilized confined or 'laboratory' populations of animals and freely growing populations in nature, so-called 'natural' populations. Confined populations in the laboratory or captive animals in the zoo are utilized to great advantage for obvious reasons. Laboratory populations can be manipulated experimentally and essential measurements are less difficult to obtain. There are, however, some serious disadvantages. One is never sure that phenomena proven to occur in a confined population would occur in an unconfined natural population of the same species. In other words, the laboratory is not a substitute for the real thing.

Light

Photic stimuli represent important external factors controlling reproduction. The manner in which this control takes place ranges from the most general direct effects of illumination on protoplasma to the psychological impact of specific visual impressions. The perception of light intensity is an intriguing mystery; the exploration of the mechanisms involved may help to explain how equally nebulous stimuli, emotional impact for example, are handled by the central nervous system.

Illumination, which means its occurrence, intensity, and the duration of the alternating periods of light and darkness on earth, is an essential component of the seasonal calendar by which most living things regulate their reproductive cycles. In translucent animals, light stimulates the germ cells directly. Thus spawning is initiated in the ascidian *Styela partita* by exposure to light.[17] Sexual maturation occurs in the hydrozoans *Hydractinia echinata* and *Pennaria tiarells* only after the germ cells have undergone a period of preparation in darkness. Maturation is then initiated by exposure to light and, once started, can proceed in darkness. Continuous light or continous darkness prevents sexual maturation and spawning. At least one hour of darkness is required, but ten seconds of subsequent illumination is enough to initiate sexual development. Isolated oocytes dissected from the gonophores, asexually produced buds, can be matured by this treatment.[18] Maturation of germ cells in *Hydractinia epiconcha* does not require alternating periods of light and darkness but takes place in direct light. However, spawning is actuated by exposure to darkness for 10 to 25 minutes.[19]

It must be assumed that substances in the germ cells respond to light or its absence and initiate a chain of reactions leading to sexual maturation and spawning in response to changes in illumination. As the gametes become shielded from light in the higher animals, the effect of light and darkness becomes by necessity an indirect one requiring mechanisms for transmitting the stimuli to the gonads. Animals with opaque skin have only a few windows such as the eyes and the parietal organs through which light can enter.[5]

Retinal photoreceptors in the eye, parietal organs (median outgrowths of the diencephalon, a region in the brain) and ocelli, as well as light-sensitive tissue in various locations of the body, may accomplish the perception of photic stimulation. Blinded minnows *(Phoxinum laevis)* are still capable of forming conditioned reflexes to photic stimuli,[20] and ducks respond with gonadal growth to increased photoperiods even though both optic nerves are sectioned or after both eyes are removed.[21] Reproductive responses are elicited by applying light directly to the hypothalamus or the pituitary gland.

Irrespective of the portal of entry, periodic light stimuli provide cues for endocrine organs which govern reproductive functions. In some species the onset of the breeding season is controlled by increasing light periods, whereas in other species diminishing light periods constitute the effective stimuli. Rowan[22] first demonstrated that additional illumination would bring the involuted gonads of the junco into reproductive activity. Ferrets are sexually dormant during autumn and winter, but they can be brought into breeding condition by keeping them in artificial light for a few hours after sunset each day.[23] Female rats receiving additional amounts of light remain in estrus for several weeks instead of the normal period of 14 hours.[24] Young female rats kept in continuous light from birth attain sexual maturity about a week earlier than normal, while those kept in constant darkness reach sexual maturity two or three weeks later than normal. Domestic sheep start to breed earlier if the amount of light to which they are normally exposed is reduced.[25]

The majority of birds and mammals have breeding periods in the spring. Such animals, woodchucks and wild trukeys for instance, are

termed long-day breeders. A minority, almost exclusively mammals
(e.g. deer, black bears), breed during the autumn; these are the short-day
breeders. Among the exceptions are salmon and trout which spawn from
October to March. It is interesting that the domestic chicken breeds the
year around, but the domesticated turkey is still a seasonal breeder.

The breeding behavior of the woodchuck is somewhat of an enigma
with respect to photoperiod. The sexually mature males emerge from
hibernation in late winter, about 30 days before the females and the
sexually immature males, at which time their testes are nearly full size.
The woodchuck testes undergo an annual cycle of enlargement and
regression with the maximum size being reached approximately two or
three weeks after hibernation. Testicular development thus occurs while
the males are hibernating in deep, dark underground chambers. The
'biological clock' for testicular recrudescence in this case may be 'set'
by the decreasing periods of daylight in the autumnal season. The
females, on the other hand, apparently respond to the increasing length
of daylight in the spring, as follicular development in the ovaries starts
soon after they emerge from hibernation.

One must wonder why male woodchucks complete hibernation a
month before the females. Tracks in the snow during later winter
reveal extensive movements and contacts with other males which
result in rather heated conflicts. I once observed three males in a
prolonged struggle in which one of the animals unfortunately lost the
end of his tail. Since the females are still underground and actual
copulation doesn't occur until much later, speculation is that the
period immediately after hibernation is a time for the males to prove
their mettle. Testosterone is responsible for aggressive behavior, and it
is probably no accident that the males emerge from hibernation with
well-developed testes. The young of the year with immature testes are
still fast asleep in their subterranean chambers.

Visual Impressions

Visual stimuli probably play an important role in vertebrate
reproductive behavior. Avian species with their elaborate courtship
rituals provide the most striking examples of the effects of such
stimuli. An isolated female pigeon is stimulated to lay eggs by the sight
of her image in a mirror.[18] Similarly, male doves separated from females
produce 'milk' for the young in their crops only if they are permitted
to see brooding female doves.[26] In male and female doves, the presence
of nesting material stimulates their readiness to incubate eggs.[27]

Odor

The single-celled protozoans and lower invertebrates are either attracted or repelled by certain chemical substances in their environment. In some instances the chemicals, either attractants or repellents, are produced and secreted by the organisms themselves. Most of the animals concerned live in a liquid environment, thus they are in close contact with the chemical elements it contains. When animals evolved on dry land, they developed olfactory capabilities to sense less concentrated chemicals. Detection and interpretation of chemicals (odors) in the environment were still required for survival. For this reason, olfactory cues are more important to the higher vertebrates than most of us realize.

The sex-attractants are well known in certain insects. Crude extracts of this material from female gypsy moths *(Porthetria dispar)* are used in survey traps to locate areas infested by these moths, destructive pests of the forest. The scent from this material attracts the male moths from long distances. A substance secreted by one individual that provokes a consistent response in another individual is called a *pheromone*. The response is automatic, sometimes involving motor reactions and often releasing courting and mating behaviour as well.

A number of glands produce odoriferous substances affecting reproductive functions. The supracaudal (anal) gland of the guinea pig, which in the male produces a waxy sebum of characteristic odor, depends on the presence of gonadal steroids.[27, 28, 29] It atrophies after castration and can be restored by injection of testosterone propionate. Administration of this hormone also promotes the development of the normally small, inactive gland of the female. Secretions from similar glands in various parts of the mammalian body serve for marking territory, as repellents, and as sex attractants. Substances in the urine, usually metabolic breakdown products of steroid hormones, may serve the same purpose. We are dealing here with amplifier systems by which the gonadal steroids increase their effectiveness as pheromones.[1]

Olfactory stimuli are also intimately related to maintenance and function of the reproductive system itself in many small rodents. The ovaries and uteri of mature female mice rendered anosmic by removal of the olfactory bulbs are significantly smaller than those of animals with intact bulbs. Corpora lutea are absent or atrophic and the accessory sexual glands are lighter and contain less secretion. No corresponding effects are observed in testes of anosmic male animals.[30]

Pseudopregnancy may be induced in intact female mice by mating them with sterilized males or by grouping them with other females, but not in subjects whose olfactory lobes have been removed.[31]

A unique response to social interaction among certain species of mice is termed the Bruce Effect. If a female mouse is removed from the stud male soon after mating and paired with a strange male belonging to a different strain from the stud male, in the majority of cases she will return to estrus and mate with the second male.[32,33] In effect the first pregnancy is blocked. Moreover, contact with the strange male is not necessary for the response; proximity is sufficient. If the female is returned to her own stud male after being separated from him for 24 hours, her pregnancy is not blocked. About 25 percent of females have blocked pregnancies when caged proximate to males of the same strain as the stud male, but if the males are of a different strain the proportion of blocked pregnancies runs between 80 and 100 percent. If the surrounding strangers are females, there is no disturbance of pregnancy.

The female is vulnerable for several days after mating. She is most susceptible within 48 hours of coitus. Resistance develops slightly by the end of the 48-hour period, but it is not fully developed until after the start of implantation between four and five days postcoitum. By the sixth day the female is virtually immune to the influence of strange males. Exposure of at least two days is necessary to elicit the response in the maximum number of females, but 12 hours is sufficient to affect about 50 percent of the females tested. Chipman and his associates[34] demonstrated that multiple, short-term exposures to strange males were also effective. Bruce reported that over 90 percent of females in which pregnancy is blocked return to estrus within four days of the start of exposure.

Here again olfactory cues are important, since anosmic females are virtually unaffected by exposure to strange males.[35] It is generally assumed that the substance or substances responsible for blocking pregnancy are associated with androgens directly (i.e. urinary excretion products of androgens) or indirectly through some androgen-dependent gland. These assumptions are based on the fact that spayed and androgenized female mice of a particular strain are as effective as normal males of that strain in blocking ova-implantation in mice of another strain.[36,37]

The immediate cause of the failure of pregnancy is failure of the fertilized ova to become implanted, probably because of the premature return of estrus and associated failure of the endometrium to become progestational. Prolactin (luteotropin) injected daily for three days after mating nearly always prevents the disruption of pregnancy. This effect of exogenous prolactin is similar to that exerted by endogenous prolactin, as attempts to block pregnancy following mating at postpartum estrus in lactating females were also unsuccessful. Attempts to obtain a similar inhibition of the pregnancy block by the administration of progesterone were much less successful, probably because this hormone substituted for the corpus luteum only during the period of injection, whereas prolactin carried the corpora lutea over the critical period so that they could continue to function later.

Subsequent studies have demonstrated that proximity of a strange male and exposure to male urine are not the only stimuli that induce the Bruce Effect. Wild house mice are extremely sensitive to the presence of a strange male, but changes in the physical environment, such as disrupting the contents of the cage twice daily, are equally effective in inducing the pregnancy block.[38] Changes in the cage size alone have been shown to reduce the pregnancy rate in deermice *(P. maniculatus)* to the same degree as the strange male.[39] Moreover, it was found that pregnancy could be blocked in deermice *(P. m. bairdii)* by exposure to females of a different subspecies *(P. m. gracilis)*.[40,41]

The effectiveness of a changing physical environment in blocking pregnancy suggested that the pituitary and adrenal glands might be part of the mechanism responsible since these glands in particular are commonly associated with responses to noxious stimuli. Parkes[42] had already suggested that excessive secretion of ACTH was probably not concerned in the reaction because this hormone did not block pregnancies when injected during the critical period prior to implantation. The fact that the incidence of pregnancy blocks was not augmented by increasing the number of males to which the female was exposed was another reason for supposing that the pituitary-adrenal axis was not implicated.[40,43]

The possible role of the pituitary-adrenal axis in the Bruce Effect was then investigated in experiments using adrenalectomized female mice from two strains of *Mus musculus,* a laboratory albino strain (CR) and a wild strain.[44] Exposure of adrenalectomized CR females to strange males of the wild strain produced no demonstrable effect but 'sham-operated' CR females with intact adrenals had a significant number of pregnancies blocked by this treatment (Table 1). Wild females were extremely sensitive to the strange males even though they were not aliens (strange males in all experimental trials were from the wild strain). Adrenalectomy again acted to protect against pregnancy block although not to the same degree as in the CR females.

In a general sense, involvement of the adrenal glands could be construed as a 'stress response'. However, in these experiments the

Table 1. Pregnancy Rates of Mice Subjected to Various Treatments[44]

Treatment and strain of female	No. of females	% Pregnant
Intact CR − no exposure	38	89
Adrenalectomized CR − no exposure	41	93
Adrenalectomized CR − exposed	35	97
Sham − CR − exposed	36	67
Adrenalectomized wild − exposed	37	49
Sham wild − exposed	42	17
Intact wild − controls	110	90

effects on reproductive function may have involved interactions between the pituitary and the adrenal cortex or the pituitary and the adrenal medulla. The action of ACTH from the anterior pituitary on reproductive function is not clear but at least the following possibilities are recognized:[45, 46] (a) direct action of corticoids on the ovaries; (b) inhibition of gonadotropin secretion by adrenal androgens; (c) direct inhibition of gonadotropin secretion by ACTH; and (d) direct action of ACTH on the reproductive tract. Other investigators have implicated epinephrine and norepinephrine from the adrenal medulla because of the potentially harmful uterine contractions produced by these hormones and the fact that adrenal demedullated mice had fewer stillbirths than controls when both were subjected to chronic stress throughout gestation.[47] Implantation could conceivably be blocked by any one of these mechanisms.

Prolactin prolongs the functional life of the corpus luteum, which in turn secretes progesterone that facilitates implantation. In this context, studies of the effects of exogenous ACTH on ovaries of mice are pertinent.[48] If the injections of ACTH were begun before sexual maturation, more than 0.5 units daily inhibited luteinization completely for up to 82 days in intact wild house mice and 0.5 units daily inhibited luteinization for 10 days. In adrenalectomized immature mice, luteinization was prevented in 16 of 18 mice for 42 days by four units of ACTH a day, but two units a day for 60 days failed to do so. Uterine and vaginal responses also indicated that ovarian function was more completely inhibited in intact than in adrenalectomized mice. ACTH was capable of inhibiting luteinization in the absence of the adrenals but inhibition was more complete when these glands were present. Daily injections of fresh male pituitary returned overian and uterine weight to normal and restored follicular maturation and luteinization in intact and adrenalectomized ACTH-treated females. The studies indicate that ACTH inhibits ovarian function indirectly through the pituitary or higher centers rather than by acting directly on the ovaries. Augmentation of the effects of ACTH by the adrenals is most likely a result of increased adrenal androgen secretion.

Sensitivity to strange males is variable in the strains of mice studied. This can be explained on the basis of inherent differences in responsiveness. Wild mice are more sensitive to changes of environment than most laboratory strains, and their levels of ACTH secretion in response to such changes are probably higher. For example, only 33 percent of the pregnancies of sham-CR females were blocked by strange males, but 83 percent of the sham-wild females' pregnancies were blocked by exposure (Table 1). Adrenalectomized wild females were not as well protected from exposure as were their CR counterparts, but this can be explained if wild house mice secrete higher levels of ACTH in response to stressors. The secretion of ACTH by some mice could be high enough to inhibit pituitary gonadotropin release directly as in the case of four versus two units of ACTH in the experiments just described.

Conclusions concerning physiological mechanisms can be summarized as follows: The pituitary-adrenal axis is involved as part of the endocrine mechanism that blocks implantation when mice are subjected to various environmental stimuli in early pregnancy. Noxious or psychologically disturbing stimuli activate a stress response which includes secretion of ACTH from the anterior pituitary. ACTH acting directly at the pituitary or higher levels to inhibit secretion of gonadotropin(s) during the critical pre-implantation stage of pregnancy is most likely a part of this mechanism. At higher secretory levels of ACTH, inhibition of gonadotropins may be acomplished in the absence of the adrenals, but at lower levels such inhibition may require the added action of adrenal androgens. Other mechanisms to explain the augmentive action of the adrenals, especially the role of medullary hormones which are also released during periods of stress, are not ruled out.

The classic stress response of Professor Hans Selye will be discussed in much greater detail with respect to intrinsic regulatory mechanisms of population growth. As far as the specificity of olfactory stimuli is concerned, it should suffice to mention that small rodents are no doubt greatly dependent on olfaction to monitor their social environment. The odor of a strange male may be an alarming stimulus to the female mouse, especially to females of the wild species. Such afferent stimuli are passed to the neural centers for integration. The neural centers in turn transmit directives to the pituitary which secretes tropic hormones. It is entirely possible that alarming stimuli, whether odoriferous or psychological, are handled in the neural centers in precisely the same fashion. A consistent response to alarming stimuli is the release of ACTH. The action of this tropic hormone as previously described could then block implantation of the fertilized ova.

Nevertheless, the possibility that a pheromone is involved in the Bruce Effect cannot be ruled out for a very good reason. Prolactin, unlike ACTH, requires a dual control system, a hypothalamic releasing factor (PRF) as well as a hypothalamic prolactin release-inhibiting factor (PIF). A pheromone exacts an automatic, consistent response, regardless of the psychological state of the individual perceiving it. The action in the integrative neural center might simply release PIF from the hypothalamus. This would block implantation without the need for additional pathways or feedbacks.

Nutrition

Reproductive failure may occur as a result of severe restriction in caloric intake or deficiencies in proteins, minerals, and vitamins. An insufficient quantity of food inhibits growth of the reproductive organs of prepubertal animals and delays the time of sexual maturity. Caloric intake and specific nutritional deficiencies affect ovulation rate, fertilization rate, and prenatal death of embryos and fetuses in domestic mammals.[49] In swine and sheep, high caloric intake increases ovulation rate but also increases embryonic mortality up to implantation time.

Reproduction can be prevented by alternately starving and feeding female mice in a series of 48-hour periods.[6, 7, 50] The embryos *in utero* are killed by this procedure but the females survive. The injection of corticosterone and ACTH can cause death of embryos in rats.[51, 52, 53] Thus, in these experiments it is possible that the stress of starvation stimulated ACTH secretion and release of lethal amounts of corticosterone from the adrenal cortex.

Restricting caloric intake in young female rats to 70 percent of the *ad libitum* levels between weaning and the day of parturition reduces mammary gland development.[54] Such restriction in mature female rats causes anestrus, irregular estrus, reduction in ovulation rate and conception rate, retardation in mammary gland development, and increased prenatal death.

Protein deficiency and low caloric intake have similar effects in cattle and swine.[49] Heifers raised on protein-deficient diets show no signs of estrus and the ovaries and uterus remain infantile. The impairment of reproductive performance due to low caloric or protein intake in domestic mammals is reversible, since adequate feeding usually restores normal function.

Certain trace elements are essential for proper growth and development of the reproductive tract and subsequent normal functioning. Deficiencies of selenium, phosphorous, calcium, manganese, copper, and cobalt are known to cause ovarian dysfunction, delayed puberty, irregular estrus, cessation of estrus, and embryonic mortality in various domestic species. Wild, free-ranging animals are no doubt similarly affected by deficiencies of these trace elements in the soil. Thus, in certain geographical locations maximum reproduction cannot be achieved.

Inadequacies in certain vitamins must also be a factor in reducing reproductive rate in natural populations. Again, there is no direct experimental evidence for this belief, but work with domestic animals has demonstrated the need for adequate levels of vitamins A, E, and B_{12} to achieve maximum reproduction.

Estrogenic substances or chemical compounds with estrogenic activity are found in several plant species. Most of these plant estrogens have relatively low potency, but they are present in certain plants in considerable amounts. The activity is usually due to isoflavones and related substances — coumestrol, genistein, biochanin A, and daidzein, for example. Reproductive failures in sheep and cattle have been associated with grazing pastures and forages with estrogenic activity. In sheep, the syndrome is characterized by dystocia, uterine prolapse, and neonatal mortality of lambs, and lactation in virgin and nonpregnant ewes.[49] A similar hyperestrogenic syndrome occurs in cattle feeding on alfalfa in Israel. Infertility, nymphomania associated with both normal and abnormal estrous cycles, swollen vulva with hyperemic mucous membranes, cystic ovaries, and hydrosalpinx are characteristics of this syndrome.[16] The physiological mechanisms of reproductive failures from plant estrogens have not been fully elucidated. They may be due to aberrations in the estrous cycle, abnormal transport of ova or spermatozoa, or failure of implantation.

Certain biologists have suggested that cyclic variations in the amounts of plant estrogens may be responsible for the cyclic variations in the reproductive rates of herbivorous mammals, especially small rodents.[55] There are no doubt a number of plant compounds with estrogenic activity still awaiting investigation. Furthermore, there may exist organic compounds with gonadotropic activity which would have profound effects on cyclic population phenomena.

Reproductive responses of predator species to variations in the numbers of their prey are worth considering from the standpoint of nutrition. When avian predators dependent on mice face a food shortage, they fail to breed. When mice and lemmings are common, these predaceous birds may raise two broods and have clutches twice the normal size.[56] A striking, well-documented example of reproductive responses to changing 'economic' conditions is found in the studies of the buzzard *(Buteo buteo)* in the British Isles.[57] Myxomatosis, a virus-induced disease that virtually destroyed the large rabbit population of the British Isles, had a secondary effect on the rabbit's specialist bird-predator. '. . . 1955 showed a great decrease in breeding activity of buzzards in all regions where rabbits had become rare or extinct. Many, perhaps most, pairs did not breed at all. It was normal, i.e. comparable with 1954, in local areas where the rabbit population was not affected, and where rabbits had never been abundant.' The buzzard of the British Isles would be recognized in America as a hawk. The name 'buzzard' in the eastern United States is applied to the turkey vulture *(Cathartes aura)* which belongs to the family Cathartidae. The same species *(Buteo buteo)* in Germany exhibits above-average fertility during years of vole irruptions (outbreaks).[58]

Pomarine jaegers *(Stercorarius pomarinus),* efficient lemming predators, were studied near Barrow, Alaska.[59] In 1951 when lemming numbers were low, jaegers did not breed at all. In 1952 there were moderate numbers of lemmings and the breeding density of jaegers averaged four pairs per square mile over the seven to nine square miles surveyed. Eighteen breeding pairs per square mile were found in 1953 when lemming numbers were high. The territory size on the most densely occupied area in 1952 was approximately 110 acres and in 1953, 15—20 acres. Nesting success was lower in 1953 than in 1952. Also, there were some breeding individuals with subadult plumage, observed in 1953, an occurrence that was not recorded in 1952 when only moderate numbers of lemmings existed.

Stevenson-Hamilton[60] has written about a similar relationship in African lions. 'So easy was it for them to catch their prey, that a lioness was accustomed to produce cubs at about twice the normal rate; in place of the usual two or three, she brought forth as many as four or five in a litter; while of these, instead of one or two only, probably all, or nearly all, were able to survive to maturity.'

The overall importance of food supply as a factor regulating or limiting population growth is not a simple proposition. The total quantity of food available for reproduction and its quality may be ultimate limiting factors, but very often population growth ceases before the food supply is exhausted. Research with laboratory populations has contributed some insight. Strecker and Emlen[61] studied the effects of a limited food supply in a confined population of wild house mice. The first response to a food shortage was noted among the younger members of the population. Before the food crisis, 89 percent of the offspring survived, but afterwards in three litters totaling 13 individuals, all but one died before reaching five weeks of age. This response was soon followed by complete cessation of reproduction. For the next six months, the biomass of mice remained more or less constant, while the actual number of mice slowly declined.

Strecker[62] followed this experiment with a study of mice in a laboratory basement where escape was possible. Food in measured quantity and water were supplied periodically. Mice were caught and marked for individual identification. A ceiling of numbers was reached and a steady population maintained, but no food shortage ever developed and there was no indication of any change in the rate of reproduction. Emigration, which had been almost negligible during the phases of population growth, suddenly increased as soon as a ceiling had been established.

These field and laboratory studies illustrate two important principles concerning nutrition and reproduction. First, an inadequate food supply is not conducive to successful reproduction. An individual, either male or female, can hardly function reproductively while starving to death. Survival of the individual takes precedence over reproduction. Second, individuals in natural populations are presented alternatives to starvation. Animals can move from one locality to another when food shortage threatens. Of course, there is the matter of competition and the possibility that food is no more abundant in another locality. Starvation occurs among the white-tailed deer in parts of North America during nearly every winter season. The fawns and yearlings, like the young mice in the populations of Strecker and Emlen, are the first to die. Either they are too small to reach the browse line or the older animals are more competitive for the limited food supply.

The problem is not so much whether nutrition influences total reproduction (for it surely does), but how food shortages affect individual members of the population and whether food supply plays an important role in regulating or limiting numbers. It is generally thought that lack of nutritional requirements has a direct adverse effect on the ovary in that raw materials necessary for the maturation of eggs and subsequent development of the zygotes are in short supply. However, there are apparently intrinsic mechanisms activated by lack of food which temporarily suspend reproductive activities.

Ovarian growth in insects is inhibited indirectly by interactions of the central nervous system and the corpus allatum. Thus, in starved insects, egg development can be initiated either by the implantation of active corpora allata or by separating the female's own corpus allatum from the inhibitory influence of the central nervous system.[63, 64] Such experiments demonstrate that lack of food is recorded through an afferent pathway. In certain species of mosquitoes, the amount of food ingested influences ovarian growth.[65, 66, 67, 68, 69, 70, 71] A distended midgut initiates the neuroendocrine chain reaction responsible for the ovarian growth. In other mosquitoes, ovarian growth is stimulated by mating or a large blood meal.[67] A species of Diptera, *Calliphora erythrocephala,* does not produce mature ova unless protein is provided in its diet.[72, 73]

Interesting parallels to these data in insects exist among vertebrates as well. Starvation induces a condition akin to hypophysectomy in tadpoles of *Rana pipiens.*[74] Gonadal atrophy and disruption of sexual cycles occur in laying pullets[75] and in rats[76, 77] when food, particularly protein, is withheld. The primary cause of gonadal dysfunction in these cases is not insufficient production of gonadotropin but failure to release it; actually the pituitary of the starved rat contains three times as much gonadotropin as that of a well-nourished rat.[78, 79] Nor are the ovaries of starved rats refractory, since they respond to the administration of chorionic gonadotropin.

There is every reason to believe, then, that the afferent stimuli resulting from starvation are routed through the central nervous system.[1] As a matter of fact, the photic stimuli provided by extra light may compensate for the effects of inadequate diet.[71] Therefore, the relationship between nutrition and reproduction is not a simple one at all; there exists the possibility that reproductive activity is curtailed in the face of impending food shortages and before the effects of inadequate nutrition are felt by the reproductive organs themselves. Neuroendocrine mechanisms for the curtailment of reproductive activity during periods of food shortages would have survival value for the individual and possibly for the population as well. Limited energy would not be wasted to reproduce young that are not likely to survive. In this case, reproduction would be sacrificed for the survival of the individual.

Population Density – Lower Limits

There are apparently lower and upper limits of density for efficient reproduction. Within limits the larger colonies of sea birds produce more eggs and hatch more young. If a colony becomes too small, there is no reproductive success, even though males and females are in contact with one another.

Naturally, the survival of the population is jeopardized when individual members become too widely dispersed. There is also a theoretical lower level of density for the population at which extinction is inevitable. In some instances, the gene pool is so reduced that interplay with environmental selective forces is no longer possible. There is also an increased chance that a precariously small number of organisms will be wiped out in an instant by lethal forces, in which case population survival becomes a numbers game. The old adage 'strength in numbers' is applicable here.

Population Density – Upper Limits

There is an optimum population density at which reproduction is maximized. As population density increases above this level, the rate of reproduction slows because of the pressures exerted on individuals by the manifold activities of other members of the population. The term 'population pressure' carries with it the idea of weight of numbers, hence pressures would be expected to increase with increased levels of density. Age composition, sex ratio, social structure, social interaction, and specific behavior patterns determine the magnitude and kinds of pressures exerted on the individual members of a population. Thus, the relation between reproductive performance and population density cannot be expressed as a simple straight line equation. Population pressures depend upon the quantity and quality of food available; the complexity of the habitat in terms of terrain, shelter, cover, visual barriers, distribution of the supplies of food and water, and climate. The whole concept of population density then becomes exceedingly complex and detailed with synergistic interactions, counterbalanced forces, constants, 'feedbacks', and weighted variables.

What chance have we to make sense out of such a complicated

system? Experimental ecologists have employed a time-proven technique that involves analysis of the component parts of the system. The secret is to measure the effects of one variable at a time, then two variables, and finally the action of three or more. In some cases an experimental population is established in the laboratory under controlled conditions. Ultimately, the ecologists move out-of-doors to gather data from natural populations. The process is tedious, time-consuming, expensive, and subject to interpretive errors. The final result is a number of theories concerning population pressures and reproductive performance which are examined in terms of mechanisms.

Social Interaction

'Social interaction' has practically become a euphemism for population pressure, mainly because the rate of such interactions is presumed to increase with increasing population density. 'Social' in the usual sense means living or disposed to live in companionship with others in the community, rather than in isolation. The adjective has been used in zoology in the strictest sense to describe animals that habitually live together in colonies or communities. Now, however, the adjective 'social' and the noun 'sociality' are being used by population ecologists to describe behavioral interactions between members of the same species.

Social interaction is not merely conflict and competitive interaction, but those behavioral responses occurring as a consequence of group interaction. Schooling, herding, flocking, dominance hierarchies, pecking orders, protection of territories, and formation of hunting packs all involve group interaction. Many of the social interactions we recognize are innate behavioral responses surfacing in group interaction. Such interactions are not in themselves harmful, but individual members of a group sometimes suffer the consequences of these interactions when the group size becomes too large. The apparent deleterious effect of intense social interaction at high population densities is the reason for the tendency to equate social interaction with adverse population pressure.

The term 'psycho-social' is often substituted for 'social' to call attention to the psychic stimuli that are reaching the brain as a result of social interaction. Social behavior is directed at other members of the species or performed in response to the actions of conspecifics. Such behavior no doubt always has psychological parameters, but more often than not the physiological consequences are slight. However, when psycho-social stimuli overload the afferent pathways to the brain, there

are certain physiological and endocrinological sequelae. The term
'social stress' is often employed to describe the manifestations of
severe emotional impact.

I have used the term 'stress' sparingly thus far because an
appreciation of the concept of stress requires a working knowledge of
neurology, physiology, and endocrinology. I have selected only the
parts of the system that are directly involved in reproduction. It is
not necessary to understand all the intricacies of the stress responses
to perceive how the system works with respect to population pressures.
I have emphasized the hypothalamus, the pituitary, and the gonads.
We need only add the adrenal cortex to the system in order to grasp
the basic principles involved. Stressors, in this case strong psycho-social
stimuli, are capable of activating the anterior pituitary to secrete ACTH.
The target organ for this tropic hormone is the adrenal cortex.
Adrenal cortical hormones secreted in response to ACTH have profound
effects on the reproductive tract. ACTH may also act directly on the
reproductive organs to affect reproduction. The results of certain
experiments with caged populations will be discussed to elucidate
many of the mechanisms involved.

In general, correlations between population density and reproductive
performance are considered *a priori* evidence for the effects of social
interaction. In most studies to date, interaction rates have not been
measured. This may appear odd and perhaps criticism is justified in
some cases. But population density *per se* has no force. The population
pressures are generated by the actions of the individual members
themselves, thus the source of the emotional or psychological stimuli
is often taken for granted.

Male Fertility in Confined Populations

The utilization of small confined groups of animals is a precise method
for determining the effects of population density and social stress on
reproductive functions. When male mice *(Mus* and *Peromyscus)* are
grouped together in small cages, certain changes in the reproductive
tracts and adrenal glands invariably occur. The adrenals increase in
size and weight, while in contrast the testes and accessory sexual glands
become smaller and weigh less than males of the same age caged alone.
These weight changes are illustrated in Table 2 from studies of male
home mice grouped 20 per cage (floor space 17.85 sq. in. per mouse).
The controls were males caged alone. The average increase in adrenal
weight after 21 weeks of groupings was 1.8 mg. or 46 percent. The
testes decreased in weight an average 29 mg. or 18 percent; the seminal

vesicles decreased an average of 64 mg. or 35 percent; and the prostate, an average of 5.6 mg. or 25 percent.

The adrenal glands respond to increased secretion of ACTH. Atrophy of reproductive organs reflects diminished secretions of gonadotropins and testicular androgens.[80] The weight changes in the adrenals and reproductive organs were accompanied by a significant decrease in the volume of semen ejaculated, in sperm densities, and in the total number of sperm per ejaculate (see Table 2).

Table 2. Mean Weights of Body, Adrenals, Testes, Seminal Vesicles and Prostate and Results of Electroejaculation Tests of Grouped and Control Males Compared (Body weights are in grams and organ weights are in milligrams)[80]

Group (N)	Body	Adren-als	Testes	Semi-nal Vesi-cles	Pros-tate	Vol-ume Ejacu-late (mm^3)	Sperm Densi-ties (1000/ mm^3)	Total Sperm (Thou sands)
CONTROLS								
A (15)	31.8	4.25	161	189	20.4	2.94	509	1480
B (13)	30.5	4.04	172	188	21.1	3.32	778	2178
C (11)	30.9	3.50	142	178	21.1	3.33	611	1960
D (11)	30.0	3.71	153	180	20.0	3.27	442	1334
E (13)	28.5	3.85	156	176	19.2	3.81	650	2441
Overall Average:	30.4	3.90	157	183	20.3	3.34	589	1886
GROUPED								
A (14)	30.7	5.57	126	134	15.6	2.99	382	836
B (11)	31.3	6.07	122	125	16.5	2.63	421	1066
C (15)	32.6	5.35	125	124	13.9	3.15	419	1000
D (11)	30.2	5.93	120	114	14.6	3.19	168	525
E (16)	30.6	5.73	141	102	13.6	3.14	790	2075
Overall Average:	31.1	5.70	128	119	14.7	3.07	477	1182

A question remains, however — are these changes followed by reduced fertility? This question was answered by checking the reproductive performance of 120 grouped males at intervals in a study lasting 60 weeks.[80] The testing procedure consisted of placing each male alone with a female mouse for two weeks in a small cage. Test females were virgins, sexually mature, and had been caged alone for at least five weeks prior to the testing. The males in this experiment were grouped for 50 weeks, and an additional 10 weeks were allotted to five breeding trials. Thus, in addition to effects of grouping on reproductive function, the factor of chronological age had to be considered. Figure 10 shows the number of mice alive at the beginning of each test period and the proportion proven fertile. Ten of the grouped mice died during the first two weeks, but losses were relatively few until 26 weeks had elapsed. During the 60-week period, only nine of the 110 control males died. Fertility was maintained in the controls until after week 26, but showed a significant drop in the grouped mice by this time. Therefore in this study at least, grouping was associated with reduced male fertility.

Altered patterns of spermatogenesis occurred along with the reduction in testicular weights of the grouped males (Fig. 11). There is considerable evidence that reduction in fertility can result from deficient sperm numbers, abnormal sperm morphology, and reduced sperm viability. However, there is no conclusive evidence that semen quality is directly responsible for implantation failures and mortality of embryos and fetuses *in utero*. Therefore, an unexpected result of the breeding trials was the significant difference between the reproductive performances of females bred to grouped and control males.

Females mated to crowded males produced smaller litters in all five breeding trials. The production of small litters was the result of a combination of factors. Females mated with controls had more ovulations, and with the exception of the first and last breeding trials also had more implantations per pregnancy. Overall, a significantly higher proportion of the embryos and fetuses of females bred to grouped males was lost *in utero*. Thus, smaller litter sizes for females mated to grouped males was attributed to fewer implantations and greater intrauterine mortality. Data on reproduction of the females from these studies are summarized in Tables 3, 4, 5, and 6.

In this study both chronological aging and crowding were found to affect fertility of male house mice.[80] The effects were manifest either

Figure 10. Survival and Fertility of Grouped Male House Mice Compared to Controls

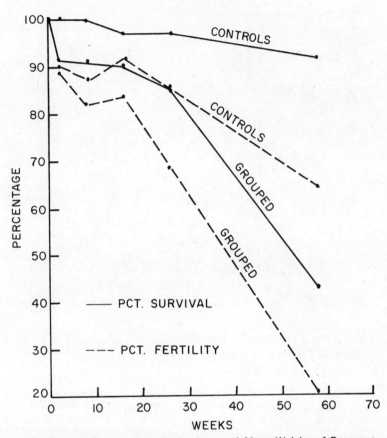

Table 3. Mean Corpora Lutea, Mean Age, and Mean Weight of Pregnant Test Females (Number of pregnant females examined are in parentheses)[80]

Weeks Grouped	Mean Corpora Lutea		Mean Age of Females in Days		Mean Weight of Females in Germs	
	Controls	Grouped	Controls	Grouped	Controls	Grouped
2	12.70 (92)	11.83 (94)	133	129	26	26
6	12.04 (92)	11.57 (84)	110	119	25	25
12	11.46 (92)	11.06 (78)	68	65	23	22
Weighted Mean:	12.07	11.49	104	104	25	24

Figure 11. Histological Features of Atrophied Testes in Grouped Male House Mice. Polynuclear Giant Cells Indicate Degeneration and Suppression of Spermatogenesis

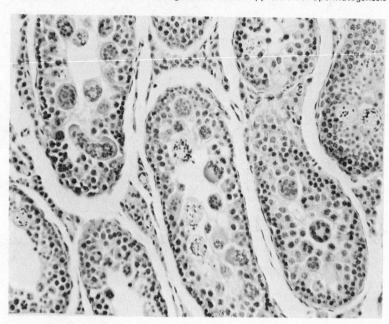

Table 4. Mean Implantations, Mean Litter Size, and Mean Number of Resorbed Embryos and Fetuses per Pregnant Test Female[80] (C = Controls, G = Grouped)

Weeks Grouped	Implantations		Litter Size		Resorptions	
	C	G	C	G	C	G
2	7.99	8.28	7.06	6.82	0.93	1.46
6	8.22	7.70	6.94	6.56	1.28	1.14
12	8.09	7.35	7.02	6.22	1.07	1.13
20	8.26	7.63	6.86	5.87	1.40	1.76
50	7.28	7.55	6.34	6.00	0.94	1.55
Weighted Means:	7.97	7.70	6.85	6.29	1.12	1.41

Table 5. Proportion of Implanted Embryos Ultimately Resorbed
(Total number of implantations given in parentheses)[80]

Weeks Grouped	Controls	Grouped
2	.1164 (791)	.1763 (811)
6	.1557 (789)	.1481 (685)
12	.1323 (793)	.1537 (669)
20	.1695 (760)	.2307 (534)
50	.1291 (473)	.2053 (83)
Average Overall	.1414(3606)	.1751(2782)

Table 6. Losses Between Ovulation and Implantation
(Number of corpora lutea are indicated in parentheses)[80]

Weeks Grouped	Mean Number Ova Lost Per Pregnancy		Proportion of Ova That Failed to Implant	
	Controls	Grouped	Controls	Grouped
2	4.71	3.55	.3709 (1168)	.3001 (1112)
6	3.82	3.87	.3173 (1108)	.3345 (972)
12	3.37	3.71	.2941 (1054)	.3354 (863)
Weighted Means:	3.96	3.71	.3274	.3233

because of primary alterations of female reproductive physiology in
response to male behavior. Social conflicts and actual internecine strife
acted as psychological stimuli to decrease secretion of gonadotropins,
which resulted in testicular atrophy and reduction of the size of the
accessory sex glands. Spermatogenesis and spermiogenesis and in turn
sperm quality and sperm concentration were reduced.

Semen quality could influence conception, embryo survival, litter
size, and even sex ratio of the offspring if vitality of the fertilizing
spermatozoa were a crucial factor in determining vitality of the
conceptus. Sex ratio would be affected if proportionately more female
determining spermatozoa occurred in the semen or if mortality were
differentially higher among male embryos and fetuses. Evidently,
there are conflicting results concerning the relation of semen quality
to fertility in humans. Certain investigators have concluded that there

is no definite evidence that any of the commonly measured semen characteristics are directly related to fertility in a cause and effect relationship. Others cite studies contradicting such conclusions and further report that, in general, poor semen predisposes to abnormalities or accidents of pregnancy. On the other hand, a considerable body of evidence from studies of lower animals indicates a definite relationship between reduced fertility and such characteristics as deficient sperm numbers, abnormal sperm morphology, and reduced sperm vitality.

The matter of sex ratio of offspring is always an intriguing problem. Data from several different sources support the general belief that intrauterine mortality in mammals takes a higher toll of the male sex. The data in Table 7 on the sex ratio of the offspring of the male house mice would support this contention. The statistical treatment of these data is included in the table, because detection of significant changes in sex ratio requires large sample sizes and mathematical proof. The data from these experiments show a statistically significant proportionate decrease in male offspring from test females bred to grouped males. The change in sex ratio is associated with increased intrauterine mortality.

Table 7. Sex Ratios of Offspring Expressed as Proportion Males (Number of offspring examined in parentheses)[80]

Weeks Grouped	Controls		Grouped		Statistical P Values
2	.5215	(650)	.5257	(622)	NS[1]
6	.5097	(616)	.4352	(540)	$P < 0.005$
12	.5285	(613)	.5000	(464)	NS
20	.5615	(618)	.4543	(372)	$P < 0.005$
50	.5095	(369)	.5200	(50)	NS
Totals:	.5276	(2866)[2]	.4829	(2048)	$P < 0.005$

Overall X^2 =   24.99, degrees of freedom = 9

1. NS = Not Significant
2. 477 (8.8 per cent) newborn mice were killed and partially eaten by mothers before they could be collected for determination of sex.

In general, the results of the experiments in question would support the argument that both fertility and litter size were related to semen quality. The higher mean number of corpora lutea found in females bred to control males was a puzzling result until studies of the effects of copulatory behavior on hormonal change in the female rat were. published by Norman Adler and his co-workers.[81] When the male rat mates, he normally delivers several intromissions of the penis before ejaculating. The stimulation provided by his copulatory behavior is necessary for the occurrence of successful pregnancy in the female. Both the transport of spermatozoa into the uterus and the hormonal conditions prerequisite to pregnancy depend on cervical stimulation by the male. Exogenous progesterone augments the percentage of pregnancies among females inseminated following a small number of intromissions. These results led to the hypothesis that multiple intromissions initiate a neuroendocrine reflex, which results in the secretion of progesterone in amounts sufficient to induce implantation and successful pregnancy. In experiments testing this hypothesis, the concentration of progesterone in systemic plasma of female rats was proportional to the amount of copulatory stimulation they received when they were in heat.[81]

Copulatory behavior was most likely an important variable in the experiments conducted with male house mice. Measurements of frequency of copulation were not available, but behavior patterns were definitely affected by grouping. The majority of the mice in the groups of 20 males assumed a subordinate position in the social hierarchy. Compared to isolated males, these subordinate animals were more submissive, less aggressive, and considerably more subdued with respect to activity patterns. Grouped animals, being less vigorous breeders, may have delivered fewer intromissions prior to ejaculating. Insufficient copulatory stimulation for progesterone secretion would explain the decreased number of implantations and the increased intrauterine mortality that were recorded for the females bred to the grouped males. Also in rats, normal coitus or comparable stimulation initiates several neuroendocrine reflexes involving the pituitary which facilitates ovulation and follicular luteinization and can lead to either successful pregnancy or pseudopregnancy. These responses are attributed to FSH and LH. Mice are apparently equally sensitive to cervical stimulation, thus after several weeks of crowding the males may not have been delivering enough intromissions to induce adequate secretion of pituitary gonadotropins. The result was a significant decrease in the number of ovulations recorded for the test females mated with crowded males. In these studies, each corpus luteum accounted for one ovulation.

Female Fertility

The reproductive processes leading to parturition are long and complicated, and malfunction at any stage is sufficient to prevent reproduction or at least influence the rate of recruitment to the population. Several aspects of female reproduction are subject to variation because of population pressures. For example, onset and duration of the breeding season, the age at sexual maturity in individual females, and the proportion of females breeding are variable and influenced by intensity of social interaction.

The effects of population pressures are manifest at all stages of reproduction, although apparently some stages are more susceptible than others. Sexual maturity in the female mammal is indicated by establishment of an estrous or menstrual cycle. In lower animals it is marked by the beginning of ovulation. In order to achieve reproductive success, the mammalian ovum must be fertilized in the oviduct (rarely fertilized once it reaches the uterus proper) and implanted in the uterus. Growth of the fetus takes place in the uterus or in the marsupial pouch. Parturition or birth of live offspring is the terminal stage of the reproductive process.

The conceptus is the zygote or any subsequently developed embryo or fetus. The loss of the conceptus can occur at any time during the reproductive cycle, but the wastage in terms of blood, tissue, and energy increases as a function of the duration of gestation. Dead embryos and fetuses of polytocous mammals are resorbed without disturbing the course of gestation of the remaining live embryos and fetuses. However, the necrotic conceptus poses a problem with respect to toxicity. The dead conceptus is easily resorbed when it is small, but female rodents are sometimes killed by toxicity when the necrotic mass is too large. Many mammals eject the dead conceptus which eliminates the problem of toxicity. Spontaneous rejection of a dead embryo or fetus from the uterus is termed abortion.

Sexual Maturity

Female rodents growing up in populations at low densities tend to become sexually mature at an earlier age than those growing up in dense populations.[82, 83, 84, 85, 86] Competition for a limited food supply may be responsible for retardation of growth and delayed sexual maturity in some instances, however, caged populations with unlimited food and water show similar responses in both males and females.[45, 48, 82, 84, 85] Since grouped animals secrete larger amounts of ACTH than isolated or paired ones, responses to endogenous ACTH were investigated to elucidate possible mechanisms. In one such experiment[45] preputial and adrenal weight of house mice increased with daily injections of ACTH and weights of thymus, spleen, ovaries, and uterus decreased. The preputial glands of mice are adjunct secretory-glands which respond to androgenic substances secreted by the adrenal cortex. At two or more units of ACTH, corpora lutea and mature follicles were absent, and follicular atresia was increased. Eight units of ACTH daily prevented maturation of the uterus.

Another experiment[48] investigated the role of the adrenal glands interacting with pituitary ACTH. Daily injections of ACTH in doses of more than 0.5 units inhibited luteinization completely for up to 85 days in intact house mice, if they were started prior to maturation. In adrenalectomized immature mice, luteinization was prevented in 16 of 18 mice for 42 days by four units per day. On this basis, delayed sexual maturity as an effect of population pressures can be attributed to increased secretion of ACTH. This hormone acting alone can inhibit secretion of gonadotropins, but its inhibitory action in the intact animal would be augmented by weak androgens secreted by the adrenals. Androgenic substances secreted by the adrenal cortex under the tropic action of ACTH are intermediate products in the biosynthesis of functional adrenal steroids. These 'weak' androgens have little effect on male secondary sex glands and male secondary characteristics, but apparently do exert rather strong feedback action on pituitary secretion of gonadotropin. Investigators[87] have demonstrated that low, nonvirilizing (non-masculinizing) doses of various weak androgenic compounds can inhibit gonadotropin secretion and sexual maturation in female mice.

The ACTH-adrenal androgen mechanism is not the only one that might delay maturation. Maternal factors (i.e. impaired lactation) and decreased secretion of growth hormone may also have significant effects. For example, Dennis Chitty[83] suggested that the population declines of the vole *(Microtus agrestis)* are due primarily to intra-specific strife at high population densities, and that the deleterious effects of such strife are passed to the offspring *in utero*, which, in turn, affects their fecundity later.

Permanent behavioral aberrations were found in young mice born to previously crowded mothers, even if they were nurtured by previously uncrowded females.[88] This suggests impairment that originated during gestation. More recently, Levine and Mullins[89] presented experimental evidence that alterations of the hormonal status of the newborn animal have profound and permanent effects on the animal's subsequent biological functioning. These investigators postulated that the action of steroid hormones in the gonadal system of the newborn animal alters the patterns and probably the amounts of hormone secretion in the adult. They also studied the effects of handling on infant rats and proposed that the increased variation of concentrations of adrenal steroids in the infant animal following handling allows the controlling set point of the brain 'hormonostat' to elicit, in the adult, graded adrenal responses appropriate to the demands of the environment. The smaller amount of variation in the non-handled animals means that, in the adult, the adrenal system will respond in a more 'all-or-none' manner. Any change in the environmental demands on the adult animal would elicit a near maximum response. In a way, early experience might be considered the conditioning stimulus that prepares the organism to cope with the contingencies of adult life.

Estrous Cycle

The established estrous rhythm is easily disrupted by social interactions involving contacts with males and other females. Although spontaneous pseudopregnancy is known to occur among female mice caged alone, its incidence increases when they are caged in small groups.[90] If the group size is increased, many of the females become anestrous.[91] The anestrous state can be abolished simply by introducing a small cage containing a single male. Also, spontaneous pseudopregnancies can be prevented by removing the olfactory bulbs.[92] The last two studies leave little doubt as to the role of olfaction in these responses. However, the endocrine pathways involved are not so clear.

The responses of mice derived from wild stock and bred in the laboratory apparently differ only in degree from those of inbred laboratory strains. Estrous cycles of wild mice caged singly or in small groups are not usually interrupted by spontaneous pseudopregnancies.[93] However, at 20 weeks of age, there is a lengthening of the estrous phase in females caged alone but not in those kept in groups. Andervont[82] described similar changes in a laboratory strain of *Mus musculus*. Estrous cycles commenced at an earlier age, were more frequent, and estrus lasted longer in segregated mice than in their matched controls kept in groups of eight.

Intrauterine Mortalility and Reproductive Rate

We have seen already how ovulation rate and implantation can be influenced by social and behavioral interactions. The embryos and festuses in the uterus during gestation are also subject to loss. Christian[84] reported an increase of 58 per cent in the number of resorbing embryos and fetuses at high density in freely growing, caged populations of mice compared to figures for controls caged as male-female pairs. Christian and Le Munyan[94] related prenatal mortality in small confined groups of mice to increased population pressures. In one experiment in which mice were crowded 20 males and 20 females to a cage for six weeks, no female became visibly pregnant. When the group size was reduced to ten males and ten females, all females became pregnant, but the number of implantations was reduced and only seven of the females had litters. Resorption of embryos and fetuses was responsible for the diminished reproductive rate. A similar relationship between group size and prenatal mortality was reported for the deer mouse.[95]

Experimental vole populations *(Microtus agrestis)* responded in a similar fashion.[96] Each young female produced an average of 10.2 young during her first year of life in a population at low density. In a second population at a much higher density the reproductive rate for young females the first year was only 1.2. In the same vein, female house mice whose reproduction had been inhibited by experimental crowding became fertile immediately after they were allowed to disperse into a larger pen.[85]

The overall study of prenatal mortality and social interaction has been sadly neglected, and very little is known about the psychological and physiological mechanisms involved. It is recognized that fertility of aged ova leads to increased embryonic mortality in the guinea pig,[97,98] rat,[99] and rabbit.[100] Changes within the gametic cell before fertilization ultimately cause the death of the developing embryo. Aging of the female gamete apparently involves physical or chemical changes with a resulting impairment of function. If changes of this nature in the ovum can lead to embryonic mortality, changes in the spermatozoan could have the same result.

The correlation between prenatal mortality and high population density is generally attributed to a 'stress response' involving the pituitary and adrenal glands. The injection of cortisol, cortisone, or ACTH in intact rats increases intrauterine mortality.[51, 52,53] This mortality can be explained on the basis of a direct action of corticoids on the developing embryos and fetuses, since cortisone at least is teratogenic.

Effects on the ovary, placenta, and uterus have not been ruled out, however. There may be several mechanisms by which ACTH affects reproductive function.[45,46] Mechanisms involving this hormone's adrenocorticotropic characteristics are dependent on the action of corticoids as outlined above or on the suppression of gonadotropin (LH) release by adrenal androgens. Additional mechanisms are independent of the adrenals, and these are postulated because of the experiments showing interference with reproductive function by exogenous ACTH in adrenalectomized mice. Christian and his students point to two possible modes of action — one whereby ACTH acts at the pituitary or a higher level to suppress release of gonadotropins and the other whereby the hormone acts directly on the reproductive tract.

Mechanisms involving the adrenal medulla cannot be ignored in a discussion of the stress response. Studies[47] showing a decreased frequency of stillbirths in adrenal demedullated mice compared to normals when both were subjected to chronic 'stress' throughout gestation implicated epinephrine and norepinephrine as potential causes of intrauterine mortality.

Laboratory Populations in General

The majority of studies of reproduction and population pressures have utilized species of mice, but studies of rabbits and chickens indicate that non-rodent species are also sensitive to crowding and confinement. High density in confined populations of rabbits has been associated with a high rate of intrauterine mortality and decreased fecundity.[101,102,103] Furthermore, the dominant females in a captive population of wild European rabbits bred much more effectively than the subordinates.[103] Pearl and Surface[104] demonstrated long ago that an increase in flock size can significantly lower egg production in chickens. Subsequently, other investigators have reported this occurrence in other flocks. More recently, George Flickinger studied social interactions in small flocks of white leghorns and described physiological responses to grouping, sex ratio, and social rank. He measured rate of social interaction by recording the number of aggressive encounters per unit time. Rate of interaction increased with group size (density or number of birds per unit of floor space was constant) and was highest when the group contained more males than females. Within the groups the largest adrenals were found among those consisting of ten cocks and two hens and the smallest among those of two cocks and ten hens. Hens caged alone or caged one hen with one cock had smaller adrenals than those in the groups. The testes of grouped cocks were significantly lighter than those of males caged alone or paired with a female. Egg production was highest in hens caged alone, and within the groups the dominant hens produced more eggs per unit time than the subordinate hens. Sexual maturity was delayed approximately two weeks when hens were placed in groups of six to twelve chickens each, and about three weeks when the sex ratio was unbalanced to favor a surplus of males.[105],[106]

Natural Populations

Inherent difficulties encountered in the field seriously hinder the collection of appropriate data for evaluating the effects of population pressures on reproductive function in nature. The most important of these difficulties is concerned with the measurements of population density and available food supply. The usual procedure has been to correlate reproductive performance with population density. Following the basic assumption that interactions between animals in the same population (intraspecific competition, social interaction, etc.) should be intensified as density increases, an inverse relationship between density and reproductive function is regarded as evidence that mechanisms similar to those discovered in the laboratory are at work in nature. The significance of these correlations is strengthened considerably if measurements of food supply are obtained simultaneously, but this has been accomplished in only a few instances.

Male Fertility (Mammals)

As one might expect, studies of male fertility in natural populations have been infrequent. Of 174 female woodchucks at least two years old examined in one locale in Pennsylvania during the breeding seasons of 1958 and 1959, I found only seven that were not pregnant. This would indicate no scarcity of fertile males. Data of this kind, however, do not indicate what proportion of males in the population were fertile. Kalela[86] found in his study of voles *(Clethrionomys rufocanus)* that attainment of sexual maturity was retarded by high population numbers. Moreover, the critical density was lower for males than for females. He concluded that the fecundity rate of juvenile voles were density-dependent and not related, directly or indirectly, to shortages of food, but rather to intraspecific intolerance arising from increasing numbers. In spite of the apparent sensitivity of juvenile voles was density-dependent and not related, directly or is improbable that this phenomenon would have much effect on population productivity. Fertility of sexually mature males is apparently not profoundly affected by social pressures in the wild, so in most situations dominant males would mate with several females. The incidence of pregnancy would not be greatly altered by the presence of a high proportion of infertile sub-adult males.

Insects

MacLogan[107] published the results of laboratory experiments with confined populations of the lucerne flea *(Smynthurus viridis)* in 1932. Above a certain density, egg production varied inversely with population density. He also noticed an individual spacing phenomenon when the insects were crowded in a glass jar. Twenty-five years later, a field study of the lucerne flea, which has become a pest of clover crops in western Australia and elsewhere, provided evidence of a density-governing reaction in the insect.[108] Initial densities were measured during May at the beginning of the winter reproductive season and final densities during September. There was a highly significant inverse correlation between initial and final densities. Similar results were recorded for four consecutive years. Changes in density patterns must have been caused by local differences in birth and survival rates, since the distances to be covered were too great for the effects to have been produced by movement in such minute flightless insects. It was possible to rule out the most obvious extrinsic factors that might have induced the changes, including soil, micro-climate, and predatory pressure.

Birds

A study of the population ecology of the great tit *(Parus major major)* in the Netherlands revealed a highly significant negative correlation between density and the number of second clutches laid.[109] This inverse correlation between population density and clutch size existed in four of five woodland areas studied. Although the density sometimes varied at a two to one magnitude, a mean difference of only about half an egg separated the higher and lower density years; none of these correlations was statistically significant. In another woodland, 'Aardenburg,' only a few nest-boxes were present from 1926 to 1929. After additional boxes were erected, the breeding population at once increased threefold. Over the next five years, the average percentage of second broods fell from 63 to 16, and the average clutch size by nearly two eggs. The average annual reproductive rate (eggs per breeding female) decreased from 12.4 at low density to 8.9 at high density. Other factors also entered into the determination of the clutch size, especially the earliness or lateness of spring and the variable productivity of the habitat from year to year.

Observations of a nesting colony of American brant *(Branta bernicla hrota)* are also pertinent. Barry[110] sampled three sections of a 4.5 mile strip of breeding area on Southampton Island in Canada. On the east end of the island where the nesting was most dense, the mean clutch size was 3.77. On the western end of the strip, the colony spread out considerably. The mean clutch size here was 4.21. On the most westerly sample area, it was 4.41.

Rodents and Lagomorphs

One of the most comprehensive field studies of the effects of population density was accomplished by a Finnish biologist, O. Kalela, on subarctic voles *(Clethrionomys rufocanus)*.[86] He perceived an inverse density effect in the proportion of young voles attaining sexual maturity and breeding during the summer in which they were born. The 1954 season started with a low population. All or almost all of the 'early-born' juveniles, males and females, became sexually mature and bred before the season ended. A high population existed in 1955. A great majority of the young males remained sexually immature, though a considerable portion of the females became fecund. In one special area where population density was about double anywhere else in 1955, practically all the young females also failed to breed. In 1956 after severe winter mortality, the population was even lower than in 1954. The majority of the early-born young of both sexes once again became mature and bred during the later part of the summer. These observations are especially pertinent because these subarctic populations of voles are cyclic. Density-dependent inhibition of sexual maturity might be a significant factor in cyclic population phenomenon.

Once the female voles studied by Kalela attained sexual maturity, they were apparently not significantly affected by population pressures, as intrauterine mortality was very low and litter size was maintained in the densest populations. Similarly Davis[111] found a higher annual incidence of pregnancy but no apparent differences in size of litters in increasing populations as opposed to stationary populations of Norway rats in Baltimore. Incidentally, he was able to increase pregnancy rates of Baltimore rats by reducing their populations.[112] On the other hand, an inverse relationship between reproduction, as measured by ovulation rate and litter size, and population density has been reported for two species of voles *(Microtus montanus* and *M. californicus).* [113]

In general, young animals are more sensitive to population pressures than older ones. The reproductive tract and perhaps the hypothalamic-pituitary axis as well are more sensitive to the inhibitory influences of stress while in the immature state. The excitatory stimuli must be greater after sexual maturity to obtain the same effects in terms of inhibition of reproduction. A corollary exists with respect to the stages of reproduction, for a greater force is required to interfere with the development of the conceptus at later than at earlier stages of pregnancy. Thus, simple inhibition of sexual maturity seems to be the usual response to moderate population pressures and such inhibition is reversible. A high incidence of 'total litter resorption' in swamp rabbits *(Sylvilagus aquaticus)* which followed sudden flooding of the habitat and resultant overcrowding is a case in point. The intrauterine losses were attributed to an 'adrenal stress syndrome'.[114] Rabbits in the wild are apparently especially susceptible to agents causing intrauterine mortality,[115] but in this case the stimuli must have been rather intense, since comparable results occur in laboratory populations only at unusually high population densities.

Formulas for Measuring Reproduction

Fecundity indicates the potential for reproduction, thus it is synonymous with sexual maturity. Fertility is a measure of the number of young produced per animal. Incidence is the number of times an event occurs during a specified period of time, while prevalence is the proportion of animals in a population that possess a specified characteristic or are affected by a certain event. For example, the prevalence of pregnant females at time t may be 50 per cent, while the mean incidence of pregnancy per fecund female during the breeding season is one.

The annual incidence of pregnancy (I_p) can be calculated from the following formula:

$$I_p = \frac{365\,P_p}{D}$$

where:

P_p = proportion of females pregnant at time t
D = duration of visible pregnancy.

The investigator would probably sample the population several times during the year, then average the data to obtain an annual prevalence of pregnancy. This technique is useful for determining the annual incidence of pregnancy in small rodent populations where a number of females can be sacrificed and dissected to examine the reproductive tracts.

A measure of fertility can also be obtained by trapping females and determining the prevalence of lactation. For example, the duration of lactation in Norway rats is 36 days. A number of females can be trapped and examined for evidence of lactation, in which case it would not be necessary to sacrifice the animals.

A number of measurements of fertility can be obtained by dissecting and examining the female reproductive tract. In many mammals, a single corpus luteum forms for each ovulation. The number of corpora lutea in both ovaries would indicate the ovulation rate. Certain species form secondary or accessory corpora lutea, thus a knowledge of the reproductive physiology of the study animal is required in every instance. Embryos and fetuses in the uterus give some indication of the average litter size of certain mammals if an adequate measure of intrauterine mortality is also obtained. An accurate measure of litter size would require observations of females at parturition. This is practically an impossibility, hence indirect methods of measuring litter size are employed instead. Implantation sites can be detected in the uteri of many mammals after parturition. These placental attachment scars remain for several weeks or even months after parturition as small brown or black spots in the uterine wall. They slowly diminish in size and eventually disappear.

In the woodchuck, which breeds only once in the spring of the year, both the corpora lutea and placental scars remain clearly visible until September, some five months after the young are born. Thus, one can measure ovulation rate, implantation rate, and potential litter size by inspecting the reproductive tracts of females killed by hunters during the summer months. Average litter size can be determined because the placental scars from resorbed embryos and fetuses, having started regressing earlier, are smaller than the implantation sites of full-term fetuses. The only error existent is the possibility of whole litters being lost *in utero*. In this occurrence, the placental scars would all be the same size and the reproductive rate calculated would then be higher than was actually the case.

The birth rate is the number of offspring per animal in the population during time t. The reproductive rate is the number of offspring per female. Specific birth and reproductive rates are calculated for various age classes in the population. For example, the annual reproductive rate of two-year-old female woodchucks in south central Pennsylvania is always higher than one-year-old females because the average litter size of the younger females is smaller and less than 50 percent of them, compared to 100 percent of the two-year-olds, are fecund. In general, reproductive rate increases with age at first, then levels out for a time, and finally diminishes with advancing age. In many mammals and birds, a stage is reached at which reproduction ceases entirely, but in nature senescence and associated loss of productivity seldom influence the reproductive force of a population.

The crude birth rate (R_b) is defined as follows:

$$R_b = \frac{B}{P}$$

where:

$\quad$ B $\ =\ $ the number of births (exclusive of stillbirths) in a given time

$\quad$ P $\ =\ $ the total living population.

Summary and Conclusions

There is a basic pattern of integrative systems involved in the management of reproductive activities which is applicable to higher invertebrates and vertebrates alike.[1] This quotation from Ernst and Berta Scharrer's book, *Neuroendocrinology*, describes it perfectly: 'It consists of a neuroendocrine control mechanism involving afferent stimuli from the internal and external milieus, their integration in neural centers, the transmission of directives to glands which furnish gonadotropins, the effects of these hormones on gonads, the hormonal feedback from the gonads to the central nervous system, and the nervous and hormonal control of accessory reproductive organs.'

A number of extrinsic agents affecting reproduction were discussed: light, visual impressions, odor, food, minerals, vitamins, chemicals, population pressures, social interaction, and certain other psycho-social factors. Food, minerals, vitamins, and certain chemicals can apparently bypass most of the integrative systems and directly affect the gonads. Estrogenic substances, for example, may act on the gonads directly and also on the neural centers and the glands that furnish gonadotropins. Components of food, minerals, and vitamins may exert their influence through their effects on general metabolism and on growth and maturation of reproductive organs. But these agents may also exert their influence through the regular afferent pathways to the neural centers. The remaining extrinsic factors affecting reproduction, although of widely diverse character, appear to operate through the same integrative systems. As a matter of fact, there are no other pathways and systems available.

Population pressures increase the intensity of afferent stimuli to the neural centers and such pressures are density-dependent. Pressures increase with increased density because of innate behavioral patterns that vary from species to species. Psycho-social factors are nebulous but exist because of intraspecific intolerance, social interaction, dominance-subordinate interactions, territoriality, and competitive interaction. At first meeting psycho-social stimuli (emotional impact) may not seem strong enough to have much effect on reproductive activity. However, once the effects of visual images and odors are demonstrated, the supposition that intense social interaction influences reproductive functions is no longer so hard to accept.

The potential role of neuroendocrine control mechanisms in regulating population density will be discussed in Chapter 5.

Notes

1. E. Scharrer, 1963. *Neuroendocrinology.* Columbia University Press, New York and London.
2. A. Gorbman and H.A. Bern, 1962. *A Textbook of Comparative Endocrinology.* John Wiley and Sons, Inc. New York, London.
3. A.V. Schally, A. Arimura, and A.J. Kastin, 'Hypothalamic regulatory hormones'. *Science* 179: pp.341-50.
4. W. Bargmann and E. Scharrer, 1951. 'The site of origin of the hormones of the posterior pituitary'. *Amer. Sci.* 39: pp.255-9.
5. E. Scharrer and B. Scharrer, 1954. 'Hormones produced by neurosecretory cells'. *Recent Progr. in Horm. Res.* 10: pp. 183-240.
6. T.J. McClure, 1958. 'Temporary nutritional stress and infertility in mice'. *Nature* 181: p.1132.
7. ——— 1961. 'Uterine pathology of temporarily-fasted pregnant mice'. *J. Comp. Pathol. Ther.* 71: p.16.
8. P. Bargeat, G. Chavancy, A. DuPont, F. Labrie, A. Arimura and A.V. Schally, 1972. 'Stimulation of adenosine 3'; 5'-cyclic monophosphate accumulation in anterior pituitary gland *in vitro* by synthetic luteinizing hormone-releasing hormone'. *Proc. Nat. Acad. Sci. U.S.A.* 69: pp.2677-81.
9. F. Labrie, N. Borden, G. Poirier, A. DeLean, 1972. 'Binding of thyrotropin-releasing hormone to plasma membranes of bovine anterior pituitary gland'. *Proc. Nat. Acad. Sci.* 69: pp.283-7.
10. A. Arimura and A.V. Schally, 1971. 'Augmentation of pituitary responsiveness to LH-releasing hormone (LH-RH) by estrogen (35249)'. *Proc. Soc. Exp. Biol. Med.* 136: pp.290-93.

11. J. Meites and C.S. Nicoll, 1966. 'Adenohypophysis; prolactin'. *Ann. Rev. Physiol.* 28: pp.57-88.
12. C.A. Nicoll, R. Fiorindo, C. McKennee and J. Parsons, 'Hypophysiotropic Hormones in the Hypothalamus: Assay and Chemistry'. *J. Meites,* ed. Williams and Wilkins, Baltimore, 1970, p.115.
13. J.E. Markee, J.W. Everett and C.H. Sawyer, 1952. 'The relationship of the nervous system to the release of gonadotropin and the regulation of the sex cycle'. *Recent Progr. Horm. Res.* 7: pp.139-63.
14. L. Debeljuk, A. Arimura and A.V. Schally, 1972. 'Effect of estradiol and progesterone on the LH release induced by LH-releasing hormone (LH-RH) in intact diestrous rats and anestrous ewes'. *P.S.E.B.M.* 139: pp.774-7.
15. C.D. Turner, 1960. *General Endocrinology.* W.B. Saunders Company, Philadelphia and London.
16. J.H. Adler and D. Trainin, 1960. 'A hyperoestrogenic syndrome in cattle'. *Refuah vet.* 17: p.115.
17. S.M. Rose, 1939. 'Embryonic induction in the Ascidia'. *Bio Bull.* 77: pp.216-31.
18. W.W. Ballard, 1942. 'The mechanism for synchronous spawning in Hydractinia and Pennaria'. *Bio. Bull.* 82: pp.329-39.
19. M. Yoshida, 1954. 'Spawning habit of *Hydractinia epiconcha,* a hydroid. Tokyo Daigaku. Rigakubu'. *J. Section IV, Zool.* 7: pp.67-78.
20. E. Scharrer, 1928. 'Die Lichtempfindlicheit blinder Elritzen (undersuchungen über das Zwischenhirn der Fische. I.)'. *Z. vergleich. Physiol.* 7: pp.1-38.
21. J. Benoit and I. Assenmacher, 1959. 'The control by visible radiations of the gonadotrophic activity of the duck hypopysis'. *Recent Prog. in Hormone Research* 15: pp.143-64.
22. W. Rowan, 1929. 'Experiments in bird migration. Manipulation of the reproductive cycle: Seasonal histological changes in the gonads'. *Proc. Boston Soc. Nat. Hist.* 39: p.151.
23. T.H. Bissonnette, 1938. 'Influence of light on the hypophysis: effect of long-continued "night lighting" on hypophysectomized female ferrets and those with optic nerve cut'. *Endocrinology* 22: pp.102-3.
24. V.M. Fiske, and R.O. Greep, 1959. 'Neurosecretory activity in rats under conditions of continuous light or darkness'. *Endocrinology* 64: p.175.
25. N.T.M. Yeates, 1949. 'The breeding season of the sheep with particular reference to its modification by artificial means using white light'. *J. Agric. Sci.* 39: p.1.
26. A.J. Marshall, 1961. 'Reproduction', in *Biology and Comparative Physiology of Birds,* A.J. Marshall, ed., Vol. II, pp.169-213. Academic Press, Inc., New York.
27. C. Champy and N. Kritch, 1929. 'Influence correlative de la castration sur les glandes odorantes et l'appareil oltaftif'. *Comp. rend. soc. biol.* 100: pp.185-7.
28. J. Martan, 1962. 'Effect of castration and androgen replacement on the supracaudal gland of the male guinea pig'. *J. Morphol.* 110: pp.285-97.
29. D. Price and J. Martan, 1961. 'The supracaudal gland of the female guinea pig'. *Am. Zoologist* 1: p.468.
30. W.K. Whitten, 1956. 'The effect of removal of the olfactory bulbs on the gonads of mice'. *J. Endocrinology* 14: pp.160-63.
31. R.L. Watterson, 1959. *Endocrines in Development.* University of Chicago Press, Chicago (Developmental Biology Conference Series).
32. H.M. Bruce, 1959. 'An exteroceptive block to pregnancy in the mouse'. *Nature* 184: p.105.

33. ——— 1960. 'Further observations on pregnancy-block in mice caused by the proximity of strange males'. *J. Reprod. Fert.* 1: pp.311-2.
34. R.K. Chipman, J.A. Holt and K.A. Fox, 1966. 'Pregnancy failure in laboratory mice after multiple short-term exposure to strange males'. *Nature* 210: p.653.
35. H.M. Bruce and D.M.V. Parrott, 1960. 'Role of olfactory sense in pregnancy-block by strange males'. *Science* 131: p.1526.
36. C.J. Dominic, 1964. 'Source of the male odour causing pregnancy block in mice'. *J. Reprod. & Fert.* 10: pp.469-72.
37. ——— 1965. 'The origin of the pheromones causing pregnancy block in mice'. *J. Reprod. & Fert.* 10: pp.469-72.
38. R.K. Chipman and K.A. Fox, 1966. 'Oestrous synchronization and pregnancy blocking in wild house mice *(Mus musculus)*'. *J. Reprod. Fert.* 12: 233-6.
39. B.E. Eleftheriou, F.H. Bronson and M.X. Zarrow, 1962. 'Interaction of olfactory and other environmental stimuli on implantation in the deermouse'. *Science* 137: p.764.
40. F.H. Bronson and B.E. Eleftheriou, 1963. 'Influence of strange males in implantation in the deermouse'. *Gen. & Comp. Endocrinology* 3: pp.515-18.
41. ——— and E.I. Garick, 1964. 'Effects of intra- and inter- specific social stimulation on implantation in deermice'. *J. Reprod. & Fert.* 8: pp.23-7.
42. A.S. Parkes, 1961. 'An olfactory block to pregnancy in mice. Part 2: Hormonal factors involved'. *Proc. 4th Internat'l Congr. on Animal Reprod.*, The Hague, pp.163-5.
43. H.M. Bruce, 1963. 'Olfactory block to pregnancy in grouped mice'. *J. Reprod. Fert.* 6: pp.451-60.
44. R.L. Snyder and N.E. Taggart, 1967. 'Effects of adrenalectomy on male-induced pregnancy block in mice'. *J. Reprod. Fert.* 14: pp.451-5.
45. J.J. Christian, 1964. 'Effect of chronic ACTH treatment on maturation of intact female mice'. *Endocrinology* 74: pp.669-79.
46. R.J. Jarrett, 1965. 'Effects and mode of action of adrenocorticotrophic hormone upon the reproductive tract of the female mouse'. *Endocrinology* 76: pp.434-40.
47. D.F. Caldwell, 1962. 'Stillbirths from adrenal demedullated mice subjected to chronic stress throughout gestation'. *J. Embryol. Exp. Morph.* 10: pp.471-5.
48. J.J. Christian, J.A. Lloyd and D.E. Davis, 1965. 'The role of endocrines in the self-regulation of mammalian populations'. *Recent Progr. in Horm Res.* 21: pp.501-78.
49. E.S.E. Hafez, 1967. 'Reproductive failure in domestic mammals'. in *Comparative Aspects of Reproductive Failure.* K. Benirschke, ed., Springer-Verlag, New York, Inc.
50. J.J. McClure, 1959. 'Temporary nutritional stress and infertility in female mice'. *J. Physiol.* 147: p.221.
51. M.C. Davis and E.J. Plotz, 1954. 'The effects of cortisone acetate on intact and adrenalized rats during pregnancy'. *Endocrinology* 54: pp.384-95.
52. J.M. Robson and A.A. Sharof, 1952. 'Effect of adrenocorticotrophic hormone (ACTH) and cortisone on pregnancy'. *J. Physiol.* 116: p.236.
53. J.T. Velardo, 1957. 'Action of adrenocorticotropin on pregnancy and litter size in rats'. *Amer. J. Physiol.* 191: pp.319-22.
54. J.F. Sykes, T.R. Wrenn and S.R. Hall, 1948. 'The effect of inanition on mammary-gland development and lactation'. *J. Nutrit.* 35: p.467.
55. G.R. Moule, A.W.H. Braden and D.R. Lamond, 1963. 'The significance of oestrogens in pasture plants in relation to animal production'. *Anim. Breed. Abstr.* 31: pp.139-57.

56. D. Lack, 1946. 'Competition for food by birds of prey'. *J. Animal Ecol.*
    15: pp.123-30.
57. N.W. Moore, 1957. 'The past and present status of the buzzard in the
    British Isles'. *Brit. Birds* 50: pp.173-97.
58. V. Wendland, 1952. 'Populationsstudien an Raubvögeln. I Zur Vermehrung
    des Maüsebussards *(Buteo b. buteo* (L.)'. *J. Arn. Lpz.* 93: pp.144-53.
59. F.A. Pitelks, P.Q. Tomich and G.W. Treichel, 1955. 'Ecological relations
    of jaegers and owls as lemming predators near Barrow, Alaska'. *Ecol. Monogr.*
    25: pp.85-117.
60. J. Stevenson-Hamilton, 1937. *South African Eden.* Cassell and Company, Ltd.,
    London.
61. R.L. Strecker and F.T. Emlen, 1953. 'Regulating mechanisms in house mouse
    populations'. *Ecology* 34: pp.375-85.
62. ———, 1954. 'Regulating mechanisms in house mouse populations: the
    effect of limited food supply on an unconfined population'. *Ecology* 35:
    pp.249-53.
63. A.S. Johansson, 1955. 'The relationship between corpora allata and
    reproductive organs in starved female Leucophaea maderae (Blattaria)'.
    Biol. Bull. 108: pp.40-44.
64. ——— 1958. 'Relation of nutrition to endocrine-reproductive functions
    in the milkweed bug *Oncopeltus fasciatus* (Dallas) *(Heteroptera: Lygaeidae)'.*
    Nytt Magasin Zool. 7: 1-132.
65. P.S. Chen, 1959. 'Studies on the protein metabolism of *Culex pipiens* L.
    III. A comparative analysis of the protein contents in the larval haemolymph
    of autogenous and anautogenous forms'. *J. Insect. Physiol.* 3: pp.335-44.
66. A.N. Clements, 1956. 'Hormonal control of ovary development in
    mosquitoes'. *J. Exp. Biol.* 33: pp.211-23.
67. J.D. Gillett, 1955. 'Behavior differences in two strains of *Aedes aegypti'.*
    Nature 176: p.124.
68. ——— 1956. 'Genetic differences affecting egg-laying in the mosquito
    Aëdes (Stegomyia) aegypti (Linnaeus)'. Ann. Trop. Med. Parasitol 50:
    pp.362-74.
69. ——— 1956. 'Initiation and promotion of ovarian development in the
    mosquito *Aëdes (Stegomyia) aegypti (Linnaeus)'. Ann. Trop. Med.*
    Parasitol. 50: pp.375-80.
70. ——— 1957. 'Variation in the time of release of the ovarian development
    hormone in *Aëdes aegypti'. Nature* 180: pp.656-7.
71. J.R. Larsen and D. Bodenstein, 1959. 'The humoral control of egg
    maturation in the mosquito'. *J. Exp. Zool.* 140: pp.343-82.
72. J. Strangways-Dixon, 1962. 'The relationships between nutrition,
    hormones, and reproduction in the blowfly *Calliphora erythrocephala*
    (Meig.). III. The corpus allatum in relation to nutrition, the ovaries,
    innervation and the corpus cardiacum'. *J. Exp. Biol.* 39: pp.293-306.
73. E. Thomsen, 1952. 'Functional significance of the neurosecretory brain
    cells and the corpus cardiacum in the female blow-fly, *Calliphora
    erythrocephala* Meig'. *J. Exp. Biol.* 29: pp.137-72.
74. S.A. D'Angelo, A.S. Gordon and H.A. Charipper, 1941. 'The role of the
    thyroid and pituitary glands in the anomalous effect of inanition on
    amphibian metamorphosis'. *J. Exp. Zool.* 87: pp.259-77.
75. T.R. Morris and Nalbandov, 1961. 'The induction of ovulation in starving
    pullets using mammalian and avian gonadotropins'. *Endocrinology* 68:
    pp. 685-97.
76. J.H. Leathem, 1959. 'Extragonadal factors in reproduction'. In *Recent
    Progress in the Endocrinology of Reproduction,* C.W. Lloyd, ed., pp.179-203.
    Academic Press, Inc., New York.

77. ———— 1961. 'Nutritional effects on endocrine secretions'. In *Sex and Internal Secretions*. Vol. I., W.C. Young, ed., pp.666-704. The Williams and Wilkins Company, Baltimore.

78. A.G.E. Pearse and L.M. Rinaldini, 1950. 'Histochemical determination of gonadotropin in the rat hypophysis'. *Brit. J. Exp. Path.* 31: pp. 540-44.

79. L.M. Rinaldini, 1949. 'Effect of chronic inanition on the gonadotrophic content of the pituitary gland'. *J. Endocrinology* 6: pp.54-62.

80. R.L. Snyder, 1966. 'Fertility and reproductive performance of grouped male mice'. In *Compatative Aspects of Reproductive Failure*, K. Benirschke, ed., pp.458-72. Springer-Verlag, New York, Inc.

81. N.T. Adler, J.A. Resko and R.W. Gay, 1970. 'The effect of copulatory behavior on hormonal change in the female rat prior to implantation'. *Physiol. and Behav.* 5: pp.1003-7.

82. H.B. Andervont, 1944. 'Influence of environment on mammary cancer in mice'. *J. Nat. Canc. Inst.* 10: pp.579-81.

83. D. Chitty, 1952. 'Mortality among voles (Microtus agrestis) at Lake Vyrnwy Montgomeryshire in 1936-9'. *Philos, Trans. Roy. Soc. Lond.* B 236: pp.505-52.

84. J.J. Christian, 1956. 'Adrenal and reproductive responses to population size in freely growing populations'. *Ecol.* 37: pp.258-73.

85. P. Crowcroft, and F.P. Rowe, 1957. 'The growth of confined colonies of the wild house mouse *(Mus. musculus L.)*'. *Proc. Zool. Soc. Lond.* 129: pp.359-70.

86. O. Kalela, 1957. 'Regulation of reproductive rate in subarctic population of the vole *(Clethrionomys rufocanus)* (SUND)'. *Annales Academiae Scientiarum Fennical,* Series A IV, 34: pp.1-60.

87. H.H. Varon and J.J. Christian, 1963. 'Effects of adrenal androgens on immature female mice'. *Endocrinology* 72: pp.210-22.

88. K. Keeley, 1962. 'Prenatal influence of behavior of offspring of crowded mice'. *Science* 135: pp.44-5.

89. S. Levine and R.F. Mullins, Jr. 1966. 'Hormonal influences on brain organization in infant rats'. *Science* 152: pp.1585-92.

90. S. Van der Lee and L.M. Boot, 1955. 'Spontaneous pseudopregnancy in mice'. *Acta Physiologica et Pharmacologica Neerlandica* 4: pp.442-4.

91. W.K. Whitten, 1959. 'Occurence of anoestrus in mice caged in groups'. *J. Endocrinology* 18: 102-7.

92. J.K. Mody, 1963. 'Structural changes in the ovaries of IF mice due to age and various other states. Demonstration of spontaneous pseudopregnancy in grouped virgins'. *Anat. Rec.* 145: pp.439-47.

93. ———— and J.J. Christian, 1962. 'Adrenals and reproductive organs of female mice kept singly, grouped, or grouped with a vasectomized male'. *J. Endocrinology* 24: pp.1-6.

94. J.J. Christian and C.D. Le Munyan, 1958. 'Adverse effects of crowding on reproduction and lactation of mice and two generations of their progeny'. *Endocrinology* 63: pp.517-29.

95. R.L. Helmreich, 1960. 'Regulation of reproductive rate by intrauterine mortality in the deermouse'. *Science* 132: p.417.

96. J.R. Clark, 1955. 'Influence of numbers on reproduction and survival in two experimental vole populations'. *Proc. Roy. Soc.,* B. 144: pp.68-85.

97. R.J. Blandau and W.C. Young, 1939. 'The effects of delayed fertilization on the development of the guinea pig ovum'. *Amer. J. Anat.* 64: pp.303-29.

98. M.C. Chang and L. Fernandez-Cano, 1958. 'Effects of delayed fertilization on the development of pronucleus and segmentation of hamster ova'. *Anat. Rec.* 132: pp.307-17.

99. R.J. Blandau and E.S. Jordan, 1941. 'The effect of delayed fertilization on the development of the rat ovum'. *Amer. J. Anat.* 68: pp.275-91.

100. M.C. Chang, 1952. 'Effects of delayed fertilization on segmenting ova, blastocysts and fetuses in rabbits'. *Fed. Proc.* 11: p.24.

101. R.M. Locley, 1961. 'Social structure and stress in the rabbit warren'. *J. Animal Ecol.* 30: pp.385-423.

102. K. Myers and W.E. Poole, 1962. 'A study of the biology of the wild rabbit, *Oryctolagus cuniculus* (L.) in confined populations. III. Reproduction'. *Australian J. Zool.* 10: pp. 225-67.

103. R. Mykytowycz, 1960. 'Social behavior of an experimental colony of wild rabbits, *Oryctolagus cuniculus* (L.). III. Second breeding season'. *C.S.I.R.O. Wildlife Research* 5: pp.1-20.

104. R. Pearl and F.M. Surface, 1909. 'A biometrical study of egg production in domestic fowl. I. Variation in annual egg production'. *Dept. Agr. Animal Ind. Bull.* 110: pp.1-80.

105. G.L. Flickinger, 1961. 'Effects of grouping on adrenals and gonads of chickens'. *Gen. and Comp. Endocrinology* 1: pp.332-40.

106. ——— 1963. 'Responses of the adrenals and gonads of chickens to social interaction'. Doctoral Dissertation, Univ. of Penna., Phila. Penna.

107. D.S. MacLogan, 1932. 'An ecological study of the "lucerne flea" *(Smynthurus viridis,* Linn.)'. *Bull. Ent. Res.* 23: pp.101-90.

108. M.M.H. Wallace, 1957. 'Field evidence of density-governing reaction in *Sminthurus viridis* (L.)'. *Nature* 180: pp.388-90.

109. H.N. Klugver, 1951. 'The population ecology of the great tit, *Parus m. major L.'. Ardea* 39: pp.1-135.

110. T.W. Barry, 1956. 'Observations of a nesting colony of American brant'. *Auk* 73: pp.193-202.

111. D.E. Davis, 1951. 'The relation between level of population and pregnancy of Norway rats'. *Ecology* 32: pp.459-61.

112. ——— 1949. 'The role of intraspecific competition in game management'. *Trans. N. Am. Wildl. Conf.* 14: pp.225-30.

113. R.S. Hoffman, 1958. 'The role of reproduction and mortality in population fluctuations of voles (Microtus)'. *Ecol. Monogr.* 28: pp.79-109.

114. C.H. Conaway, T.S. Baskett and J.E. Toll, 1960. 'Embryo resorption in the swamp rabbit'. *J. Wildl. Mgt.* 24: pp.197-202.

115. H.G. Lloyd, 1970. 'Variation and adaptation in reproductive performance'. In *Variations in Mammalian Populations.* R.J. Berry and H.N. Southern, eds. Symp. Zool. Soc., London, No. 26, pp.165-88.

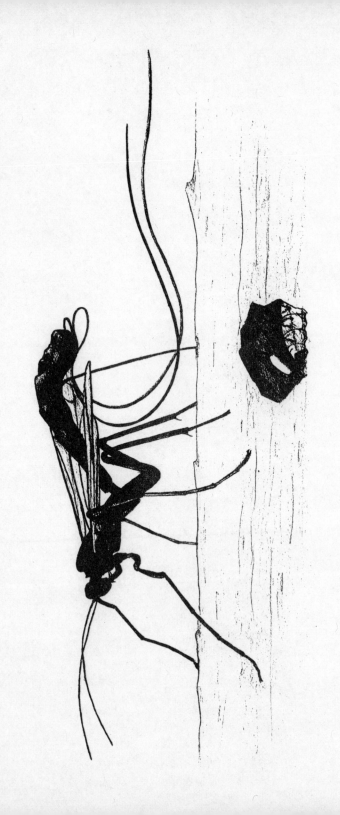

3 POPULATION DECLINE – MORTALITY

Mortality, measured as the number of individuals dying during a specified period of time, is the relative frequency of death and the force for population decline. The causes of death, which are legion, and the patterns of mortality in various populations are the principal subjects of this chapter.

Organisms die from a great variety of causes: accidents, predation, parasitism, malnutrition, starvation, inclement weather, disease, and 'old age' or senescence. Any measure of the probability of dying during a particular age interval (mortality) or the probability of surviving to a particular age (survivorship) is a population characteristic.

Accidental Deaths

As if the hazards of the natural world were not enough, wildlife today must avoid a number of man-made technological contrivances to escape traumatic death or injury. Automobiles, trucks, buses, trains, airplanes, and watercraft annually take their toll of animal life. It may be hard to believe, but 26,445 white-tailed deer were killed by motor vehicles on Pennsylvania highways during 1974 alone. This total was more than the combined number killed by hunters in 35 other states during the same year. I counted 53 animals dead on the roadways of a 6,820-acre study area in western Pennsylvania during a one-year period: 36 cottontail rabbits, nine opossums, two woodchucks, one gray squirrel, one skunk *(Mephitis mephitis)*, one muskrat *(Ondatra zibethica)*, one weasel, one ring-necked pheasant *(Phasianus colchicus)*, and one bobwhite quail *(Colinus virginianus)*. Many others have made similar tallies in their favorite wildlife habitats and obtained equally appalling figures. Fires started by lightning or accidentally by man frequently decimate vast areas of both plant and animal life. Power lines, hydroelectric turbines, dams, sluice gates, and plate glass windows are other lethal devices that make for a precarious existence.

A well-documented story of the fate of Pacific salmon *(Oncorhynchus* sp.) in the Fraser River system in Canada's easter British Columbia illustrates how oblivious man can be of the simple needs of valuable wildlife species. Mature salmon congregate in the Johnstone Strait and the Juan de Fuca Strait at the northern and southern ends of Vancouver Island before making the spawning run up the inland rivers and streams. In 1913 approximately 31 million returned from the Pacific Ocean, of which 24 million were taken in nets, leaving 7 million to make their way to the spawning grounds.

In one part of Canada, the Fraser River flows through one of the deepest canyons in North America. The tracks of the Canadian Pacific Railroad were completed on the west bank in 1885. The Canadian Northern was working on a transcontinental line along the east bank in 1913. Workmen dumped tons of rubble and rock blasted from the cliff walls into the turbulent stream without giving a second thought to the needs of migrating fish. In 1913, vast numbers of salmon were blocked at a point below a place called Hell's Gate. Consequently the Fraser and its side channels literally turned red with dead and dying fish. The Fraser's commercial salmon yield dropped from 730,000 cases in 1913 to less than half that amount in 1914 when rock slides increased the obstruction even further.

Chinook salmon *(O. tschawytscha)*, sockeye salmon *(O. nerka)* and silver salmon *(O. kisutch)* spend the first two to three years of their lives in inland rivers and streams. A few males reach sexual maturity while still in fresh water, which incidentally happens with the Atlantic salmon as well. Dog salmon *(O. keta)* and pink salmon *(O. gorbuscha)* have a slightly different life cycle. The young begin migrating toward the sea just a few weeks after leaving the nest when they are only three or four centimeters long. Chinooks reach sexual maturity at an age of four to seven years at weights of 25-45 kg. The pink salmon, one of the smaller species with a weight range of two to three kilograms at sexual maturity, has the shortest life span — two years. With the exception of a few scattered males, all species of salmon complete their growth in salt water. The failure of salmon to spawn in one season would not be felt in commercial runs until several seasons later, thus the disastrous effects of habitat destruction may not have been immediately discernible. This is generally true in the majority of cases. The general public becomes aware of the problem only when the damage has already been done. Thus in the Fraser River system the spawning runs had probably already been affected even before the fateful years of 1913 and 1914.

Not quite thirty years ago, ingenious fishways were completed at Hell's Gate at a cost of $1,000,000. But by then, according to W.F. Thompson, first Director of the International Pacific Salmon Fisheries Commission, US and Canadian fishermen had been deprived of catches worth some $330,000,000 or $12 — $15,000,000 annually.

Predation and Parasitism

Trauma is a frequent cause of death in nature. Mortal injuries
inflicted by conspecifics are uncommon. Predation and parasitism
in which the damage is done by another species are responsible for
most of the deaths caused by trauma.

A predator is an animal or plant that catches animals and eats
them. The difference between predation and parasitism is not very
great, yet most of us recognize the distinction between the two
activities. The predator is usually larger than its prey, although
the exceptions, timber wolves preying on moose for instance, are
fascinating. A parasite must be smaller than its host. Any other size
ratio would literally drain the life from the parasitized organism.
The prey is always an animal. When an animal catches and devours a
plant, the act is not termed predation. Plants are considered
predaceous if they trap and consume insects and other small
invertebrates, but when they feed off other plants, they are deemed
parasitic. There are similarities but everyone appreciates the
differences, real or fancied. Perhaps the best way to tell a predator
from a parasite is by the results. A predator always kills the prey. The
strategy of the parasite is to allow its host to live. When an animal
kills and eats one of its own species, the deed is pronounced cannibalism.
Parasites apparently do not infect conspecifics, so there is no word to
describe such a condition.

Predator-prey interactions with respect to population dynamics
have received considerable attention from ecologists. Prey populations
largely determine the growth rate of predator populations because
they provide the nutritive resources for growth, maintenance, and
reproduction. We have already seen how the reproductive rates of
predaceous birds and mammals are influenced by the abundance of
their prey. When northern murids (lemmings, mice, and voles) are in
short supply, foxes starve to death in large numbers and snowy owls
(Nyctea nyctea) migrate south into the United States, sometimes
even as far as North Carolina, in search of prey. This irruptive
migration is apparently a one-way movement, since few if any of
these owls ever return. Thus, the shortage of prey influences two
other population forces of the predators, mortality and dispersal.

Predators tend to reduce the growth rate of prey populations,
but whether they very often substantially reduce the equilibrium size
of the prey population is a controversial question. Certain studies
demonstrate relatively effective control of prey densities in certain
situations, while other investigations find only minimal control at best.

There are apparently many interacting phenomena in each community that determine ultimately whether predators at any given moment in time exert control over the growth rate of the prey population.

Experimental studies of the relationship between cyclamen mites *(Tarsonemus pallidus),* an economically important pest of strawberries in California, and predatory mites of the genus *Typhlordromus* indicate that predation can be an important regulatory factor in arthropod populations.[1] Cyclamen mites typically invade a strawberry patch the first year it is planted, but populations do not build up to high densities until the second year, when they can cause heavy damage. The predatory mites usually invade the patches during the second year and rapidly decimate the pest populations. Outbreaks of the cyclamen mite usually do not occur again once the predators are established.

In field plantings of strawberries, the insecticide parathion was used to control insect pests. Apparently parathion kills the predatory mite but has no effect on the cyclamen mite. Thus, populations of the pest mite reached epidemic levels where the insecticide was used. When cyclamen mites began to build up in untreated patches, the predator mites quickly moved in to quell the increase. Overall, cyclamen mites were about 25 times more abundant in the absence of predation.

The effectiveness of *Typhlodromus* in controlling prey densities is not determined by its voraciousness alone. The reproductive rate of the predatory mites is about equal to that of the cyclamen mites. Both species reproduce parthenogenetically. Female cyclamen mites lay approximately three eggs per day for the four to five days of their reproductive lifespan; female *Typhlodromus* lay two or three eggs per day for eight or ten days. The reproductive capacity of most predators is well below that of their prey. However, high reproductive rate alone is not the primary reason for the efficiency of *Typhlodromus* as a predator. Synchrony of seasonal reproduction with the growth of the prey populations, ability to persist at low prey densities, and strong dispersal powers are important elements of predator efficiency. During the winter months when the pest mites exist in small numbers safely hidden in the crevices and folds of the leaves in the crowns of the strawberry plants, the predatory mites subsist on the honeydew secreted by aphids and white flies. The female predatory mites reproduce only when they are feeding on other mites.

In nature a prey species is seldom preyed upon by a single predator species even for a short period of time. In the northeastern region of the United States, the following predators have been known to prey on ruffed grouse: red fox *(Vulpes vulpes)*, gray fox *(Urocyon cinereoargenteus)*, New York weasel *(Mustela noveboracensis)*, Bonaparte weasel *(M. cicognanii)*, skunk, raccoon, bobcat, opossum, woodchuck, red squirrel *(Sciurus hudsonicus)*, chipmunk *(Tamias striatus)*, various species of mice and shrews, mink *(Mustela vison)*, marten *(Martes americana)*, fisher *(M. pennanti)*, lynx *(Lynx canadensis)*, porcupine, dog, housecat, great horned owl, barred owl *(Strix varia)*, goshawk *(Accipter atricapillus)*, Cooper's hawk *(A. cooperii)*, sharp-shinned hawk *(A. velox)*, raven *(Corvus corax)*, crow *(C. brachyrhynchos)*, duck hawk *(Falco peregrinus)*, red-tailed hawk *(Buteo borealis)*, marsh hawk *(Circus hudsonius)*, broad-winged hawk *(Buteo platypterus)*, red-shouldered hawk *(B. lineatus)*, osprey *(Pandion haliaetus)*, snowy owl, and screech owl *(Otus asio)*. To this list can be added the ring-necked pheasant which occasionally causes nesting failure by laying eggs in grouse nests and the northern blacksnake *(Coluber constrictor)* which has been caught in the act of nest-robbing.

This formidable array of 'enemies' would greatly alarm an ardent grouse hunter or an admirer of this splendid bird of the forest. However, considered in proper perspective the list is not nearly so ominous. First of all, one must remember that the newly hatched grouse, weighing only 13 grams, is not as large as many of the mice and shrews in its native habitat. Many of the predaceous birds and mammals listed as grouse predators are specialized predators of small rodents, but would be expected to include a grouse chick or two in their diet if the opportunity presented itself.

This brings us to the important principles of relative densities of prey species, 'buffer' species, expenditure of energy, and the point of diminishing returns. The broad-winged soaring buteo hawks could hardly make a living out of preying on grouse chicks because of their low abundance and the short period of time during the year they exist. The larger hawks and owls thus are adapted to hunt the abundant small mammals that exist the year around. If they kill a small grouse, it is more accident than design. The Cooper's hawk and the sharp-shinned hawk, on the other hand, are specialized bird predators and more of a threat to grouse of all ages. But these too concentrate largely on other avian species, which although smaller than grouse in the aggregate occur at greater population densities.

Game managers apply the term 'buffer' to those species that serve as the staple food of predators and thereby lessen the pressure on game animals. In order to serve in such a capacity, a prey species must be present in far greater numbers than the game species and be relative ly easy to catch. The principal buffers with respect to the ruffed grouse are small woodland birds, the cottontail rabbit, varying hare (snowshoe hare), squirrels, mice, and shrews. Since the cottontail rabbits, hares, and squirrels are also considered desirable game animals, the application of the term 'buffer' depends on which animal is considered the object of value.

The authors of *The Ruffed Grouse,* published by the New York State Conservation Department, were convinced after an investigation lasting 13 years that losses of grouse from predation were largely concerned with surpluses.[2] If grouse had not been killed by predators, they would have succumbed for some other reason. 'The species seems to have an aversion to densities above a bird to about four acres of cover and tends under such circumstances to disperse to less occupied territory.' The New York investigators concluded that a reduction in predators could not be expected to bring about any lasting increase in a grouse population.

Nevertheless, the recorded losses from predation were considerable. Beginning with the breeding season such losses may be logically divided into three categories corresponding to the major life periods of the grouse: nest, brood and adult. Nesting failure represents the first major loss during most years. From 1930 through 1942 a total of 1,431 nests were examined by the New York State Grouse Investigation. In each case the fate of the nest was noted and a diagnosis of the cause was made for those that had been broken up. Table 8 presents the results with respect to total losses and the proportion attributable to predation. Over the areas surveyed, nest destruction averaged close to 39 percent and of those nests destroyed 89 percent were attributed to predation. Foxes were apparently responsible for a major share of the nest destruction.

The brood period defined by the Grouse Investigation extended from hatching to 31 August. Throughout the thirteen years of the study, the total brood mortality was measured each summer on two major study areas in New York. On both, its course was remarkably similar, averaging slightly over 60 percent by the end of August (Table 9). Although many chicks were consumed by predators, available evidence indicated that predation was not the cause of these early season losses even though in many cases a predator may have been the final executioner. Confined chicks hatched artificially from eggs collected in the wild suffered heavy natural losses in June which were unrelated to predation. Nearly half of the mortality among wild broods

Table 8. Grouse Nest Mortality Recorded in New York State and Proportion Resulting From Predation 1930−1942[2]

Nest data	Year													
	1930	1931	1932	1933	1934	1935	1936	1937	1938	1939	1940	1941	1942	Total
Total nests*	13	206	211	216	67	61	74	72	121	145	150	72	23	1,431
Number hatched	5	126	124	125	41	39	54	55	77	101	72	48	11	878
Number broken up	8	80	87	91	26	22	20	17	44	44	78	24	12	553
Per cent broken up	61.1	38.8	41.2	42.1	38.8	36.0	27.0	23.6	36.3	30.3	52.0	33.3	52.2	38.6
Per cent of broken-up nests attributable to predation	62.5	82.9	74.7	87.8	92.0	100.0	78.2	94.1	90.9	97.7	96.3	100.0	91.7	89.0

*Includes only nests for which survival data are comparable.

Table 9. Brood Mortality Recorded on Connecticut Hill and Adirondack Study Areas − 1930−1942[2]

Area	Year													
	1930	1931	1932	1933	1934	1935	1936	1937	1938	1939	1940	1941	1942	Total
Connecticut Hill	57.5	70.0	54.8	76.7	51.4	80.8	54.2	55.9	62.4	63.0	57.9	59.5	77.6	63.2
Adirondack	....	.....	58.3	78.4	50.8	88.3	41.8	60.0	72.0	57.0	57.0	64.0	55.5	60.9

occurred by July 1.

Losses experienced during the rest of the summer appeared more directly attributable to predator activity. The principal predators at this age were Cooper's and sharp-shinned hawks. Considering all the data collected, apparently upwards of one-third of the chicks hatched succumbed as a result of predation. The proportion represented a greater number of individuals during those years when the hatch was higher. The percentage mortality, however, was not proportionate to the size of the hatch, which means that predator pressure on young grouse was density-independent.

Grouse grow to adult size between 1 June and 31 August. The New York Investigation measured adult mortality between September and the following April, the commencement of the next breeding season, on the two special study areas. On Connecticut Hill losses during this period averaged 42 percent from 1930 to 1941 and in the Adirondack area, 53 percent during the same period. Annual mortalities and the initial September populations on the two study areas from 1930 to 1941 are listed in Table 10. The annual mortality was measured from 1 September of one year to 31 August of the following year. Mortality rates were not related to the initial September populations, thus adult mortality was also density-independent. The investigation concluded that predation was responsible for over 80 percent of the annual mortality among adult grouse, or close to 40 percent of the fall population each year. Accidents, disease, and other circumstances made up the remaining 15 or 20 percent of the annual mortality.

The importance of ruffed grouse as a game bird has led to several predator control experiments. A total of 557 predatory birds and mammals were removed from about 2,000 acres in another New York study area, and an adjacent area left uncontrolled as a reference. The data from this experiment are tabulated in Table 11. Predator removal improved nesting success, but there was no carry-over to higher population densities in the fall. After a year for recovery, this experiment was repeated with the reference and experimental areas reversed. The same results were obtained; predator control reduced nest losses but did not influence chick mortality.

The experiment was repeated on an island population of grouse later.[3] From 1940 to 1945 predators were removed from 1050-acre Valcour Island in Lake Champlain, New York. The results were again the same. Predator control increased nest success but had little effect on chick losses or adult mortality. A substantial decline in

Table 10. Adult Mortality Recorded on Connecticut Hill and Adirondack Study Areas — 1930—1941*[2]

Area	Adult grouse data	Year												
		1930	1931	1932	1933	1934	1935	1936	1937	1938	1939	1940	1941	Aver.
Connecticut Hill	Initial September population	161	276	465	274	420	300	311	273	334	394	276	300	315
	Per cent mortality	14.3	28.9	57.6	40.5	52.6	52.0	63.0	42.8†	51.2	58.8	54.7	57.7	50.2
Adirondack	Initial September population	....	60†	70	51	73	23	64	59	42	70	63	53	57
	Per cent mortality	....	46.7	62.9	43.1††	78.1	30.4	71.9	57.6	28.6	65.7	60.3	56.6	57.9

* Year indicated is that of the beginning of each 1 September to 31 August period.

† Estimate made in January 1932 and therefore somewhat below September level.

†† In 1933 on the Adirondack area and in 1937 on Connecticut Hill, mortalities of 52.9 per cent and 45.8 per cent respectively were recorded by April but during the ensuing summers more birds moved into the areas than were lost.

Table 11. Results of Predator Control on Grouse Populations[4]

| | 1931 | | 1932 | |
	Predators Removed	No Removal	Predators Removed	No Removal
Nest loss (percentage)	24	51	39	72
Chick mortality (percentage)	57	67	54	55
Adult loss (percentage)	11	15	32	21
Grouse population density in fall (birds per 100 acres)	13.0	9.8	18.7	18.0

population density of grouse took place over the winter of 1943-44 on Valcour Island in spite of the intensive predator control program.

Small rodents, especially voles and lemmings are subjected to intensive predation by birds and mammals. The impact of mammalian predators on high vole populations before and during predator control programs was measured on a 35-acre study plot. The results are presented in Table 12. A smaller percentage of the voles was killed by predators in 1963 during predator control, but the population crash occurred as before. Predators again appeared to be feeding on a doomed surplus of prey. The greater mortality must be assigned to another cause.

Table 12. Impact of Predation on a Population of the California Vole *(M. californicus)*[8]

	1961 Before Predator Control	1963 During Predator Control
Population decline of voles (July–January)	4,400 → 100	7,600 → 200
No. voles killed by carnivores during this decline	3,870	1,916
Destroyed by carnivores (percentage)	88	25

Less than a century ago, the timber wolf *(Canis lupus)* occurred throughout North America, but today it is absent from 45 of the 48 contiguous states. Probably less than 500 individuals inhabit the

remaining three — Minnesota, Wisconsin, and Michigan.[5] Whether this
predator is capable of controlling populations of deer, moose *(Alces
alces)*, caribou, and wild sheep is apparently dependent on the ratio
of prey to predators. Murie[6] in 1944 concluded that wolves controlled
Dall sheep *(Ovis dalli)* in Mount McKinley National Park in Alaska
where there were approximately 25 to 37 sheep per wolf. In areas where
wolves do not control prey populations, the ratio is much larger.
In the Rocky Mountain national parks of Canada, there were 300 to
400 head of big game per wolf in 1947.[7] Cowan concluded from his
studies in the area that all the predators present, coyote *(Canis latrans)*,
wolf, fox, lynx, wolverine *(Gulo luscus)*, mountain lion *(Felis concolor)*,
grizzly *(Ursus horribilis)* and black bears, were not taking the annual
net increment to the game herd, nor even removing the 'cull group'. A
cull is a weak or inferior animal that would ultimately die from
disease, parasitism, or malnutrition. Such animals are normally
'culled' or removed by predators. Another wildlife biologist, Stenlund,[9]
estimated in 1955 that wolves were killing about 16 percent of the herd,
much less than the annual turnover, in areas of Minnesota where there
were about 153 deer per wolf.

In general, predator control programs in most of the areas
investigated had probably greatly reduced the wolf populations.
David Mech studied predator-prey interactions on Isle Royale in
Lake Superior where both moose and timber wolves are protected
by law.[5] Mech's remarks about wolf-moose relations are based on
studies conducted between June 1958 and June 1961. The important
question is whether wolves merely substitute for other mortality or
whether they kill more animals than other factors would. The
history of the Isle Royale moose herd affords an answer in part.
Before wolves became established, the herd increased to an estimated
1,000 to 3,000 animals in the 1930s, decreased drastically a few years
later, and built up again in the late 1940s. 'The limiting factor was
food.' Signs of severe overbrowsing are still evident. Apparently the
wolves are maintaining the moose population below the level at
which food would restrict it (1961). The prey-predator ratio on Isle
Royale was 30 moose per wolf.

An obvious result of intensive predation on Isle Royale moose
was the elimination of heavily parasitized, diseased, old, or otherwise
inferior individuals.[5] That predators are more likely to catch the
weakest, the slowest, the less alert prey is accepted generally, and there
is probably some truth to this supposition. Culling supposedly
benefits the population by removing inferior animals. The benefit would

accrue from eliminating inferior genes from the gene pool and leaving more food for the surviving members of the population.

David Mech, writing in the *American Journal of Natural History* in 1975 about subsequent wolf studies on Isle Royale, pointed out that moose are increasing again and that the wolves are in fact now cropping only part of the surplus. Will they have more effect when the moose begin to outstrip their food supply and weaker ones become easier prey? Will the wolf packs become larger and more numerous? Only further research will answer these important questions.

Predator control does not always produce the desired results. During the spring of 1974 I visited certain wild areas in southern Africa where this fact was demonstrated explicitly. In many parts of South Africa, the graceful impala *(Aepyceros melampus)* is the most numerous antelope, but this was not always true. Impalas are highly gregarious. Each dominant buck has a harem of several ewes, which often numbered 50–100 in the areas I visited. Males, young and old, form large bachelor herds. Their movements are very fast, and they are capable of prodigious leaps of possibly 20 or 30 feet. Their principal enemies are the leopard *(Panthera pardus),* the cheetah *(Acinonyx jubatus),* the wild dog *(Lycaon pictus),* and occasionally the lion *(Panthera leo).* Biologists in southern Africa attribute the growing populations of impalas to the fact that these large predators have been eliminated needlessly in many regions. The small, golden antelopes breed at a great rate and eat the grass down to the ground, thereby destroying food and cover for all the other antelope species, large and small. The impalas thus have become disproportionately abundant relative to the other herbivores in southern Africa.

The matter of relative densities of herbivores in the face of predator pressures is confirmed in the wilderness areas of Botswana where the elimination of the large predators has not proceeded to the same degree as in most of southern Africa. The bush country of Botswana, largely uninhabited by man, supports a number of unusual species of antelope, e.g., sable antelope *(Hippotragus niger),* red lechwe *(Kobus (Adenota) leche leche),* and the sassaby or tsessebe *(Damaliscus lunatus).* Impalas were not nearly as plentiful as in the other parts of southern Africa I visited. The reason for this was brought to mind forcefully one day in the field when I observed a pack of 20 wild dogs (or hunting dogs) dispatch one of these speedsters. The pack followed a standard practice of the wild dogs. Two dogs pursued the single impala closely until it accidentally circled back to run into the main pack running parallel to the ones in pursuit. The method of pursuit observed at close hand and on foot was so efficient that I wondered at the time if it would have mattered very much whether the impala selected for the chase was inferior or not. The wild dogs have been eliminated from large areas of southern Africa, and such a large pack is no doubt a measure of the level of 'predator control' practiced in Botswana.

Insect parasitism is apparently an effective mechanism for controlling populations because the infective rate is density-dependent. Data collected by Varley[10] on a major parasitic insect *(Eurytoma curta)* illustrate this relationship in Table 13. In general, a relatively greater number of the host can be located and attacked at high population densities because many of the victims are being forced into more exposed positions. Introduced insect pests quickly reach epidemic proportions because there are few if any natural parasites in the new habitats. Effective control of an introduced pest species normally requires the importation of its natural enemies.

Table 13. Larvae of the Knapweed Gall-Fly *(Urophora jaceana)* Parasitized by the Parasitic Insect *(Eurytoma curta)*[10]

Year and Population Level	Larval Population at Beginning of Season (no./m^2)	Larvae Killed by Parasite (no./m^2)	Per Cent Population Killed by Parasite
1934 (low population)	43	6	14
1935 (high population)	148	66	45

An unusual synergistic relationship exists between a parasite and a predator in North America. The adult tapeworm *(Echinococcus granulosa)* lives in the intestines of the wolf where it causes little harm. Its ova are passed to the external environment in the wolf's feces and contaminate the vegetation and water. The ova are ingested by herbivores, some of which are the wolf's prey, and hatch in the duodenums of these intermediate hosts. The oncospheres (newly hatched embryos) migrate through the intestinal wall, enter the mesenteric venules and become lodged in the capillaries of various organs and tissue including the liver, lungs, heart, and spleen. The developing larvae produce the hydatid cyst which expands slowly but continuously during the life of the host animal. From the inner germinal layer of the hydatid brood capsules develop which may remain attached or set free into the fluid of the cystic cavity. Scolices develop from the inner wall of these capsules. Each of the scolices in a hydatid becomes an adult worm in the intestines of a suitable carnivorous predator. The life cycle continues as before.

The tapeworm larvae of *E. granulosa* are found in 30 to 68 percent of the moose examined in North America.[5] This species of tapeworm is especially virulent as a pathogen because the larvae multiply within the vital organs of the intermediate host. The damage produced by the hydatid cyst of *E. granulosa* is both mechanical and toxic. It will eventually cause debility, diminished resistance to secondary infections and death. Thus, the wolf is the unwitting host to a parasite that increases the efficiency of its predation on moose populations. It spreads the agents of the prey's eventual destruction throughout the moose's range, thus assuring that eventually the parasitized victims will become weak enough to be chased down and killed. Coyotes and foxes that feed on moose carcasses also help to spread the parasite and maintain its population. Foxes are primary hosts of *E. granulosa* in Europe.[11]

Malnutrition and Starvation

Nutritional diseases as causes of death in nature have been largely unexplored. Starvation from inferior quantities of food is considered an important cause of mortality by many wildlife biologists, but the quality of food in terms of protein levels and essential vitamins and minerals has received scant attention. A belief in the inherent adequacies of nature is apparently a prevailing attitude. Yet there is reason to believe the contrary. Animals in nature suffer from mineral and vitamin deficiencies and an adequate level of protein may very well be a perennial problem.

Malnutrition and starvation caused spectacular losses of deer in North America during the first half of this century. During the 1800s, deer herds in the eastern United States were limited by market hunting, predators, natural losses, and food restrictions in the mature forests. Deer almost vanished from the State of Pennsylvania around 1900, with only about 200 remaining. Between 1900 and 1925, measures were adopted to protect deer and 1,200 animals were stocked. The population mushroomed to an estimated 1,000,000 animals in 1939. The female deer were protected until 1923, when antlerless seasons were initiated to obtain heavier harvests and reduce winter losses from starvation.

The perennial effects of malnutrition and starvation in the large deer populations in Pennsylvania were the impetus for excellent research studies conducted by the Pennsylvania State University and the Pennsylvania Game Commission. A 150-pound buck needs about 10 to 12 pounds of good quality natural food for best growth and antler development. Male fawns have only 'buttons' during the fall at six months of age, but the same deer at 1½ years of age may have up to 10 or 12 points or tines on their antlers. This may come as a surprise to hunters in Pennsylvania and elsewhere who are accustomed to bagging deer with only spikes or small branched antlers. In growing bucks, body growth takes priority over antler development. This can account for deer being in good condition yet lacking the extra nutritive materials to develop large, branched antlers. On restricted diets, either in quantity or quality, bucks may have buttons or spikes an inch or so long at 1½ years, and if kept on restricted diets, they may never develop branched antlers. On a complete diet, bucks may have from 4 to 12 points at 1½ years.

For best growth, deer need a diet containing 13 to 16 percent protein. Calcium and phosphorous needs are interrelated. Deer on low energy, low calcium and phosphorous, and low protein experimental diets will be stunted in size and weight and will develop thin spike antlers. In comparison, deer on 'complete' experimental diets have large, heavy antlers with 6 to 14 points.

Measurement of antler size and development in the field should reflect food conditions. For example, according to wildlife biologists, Potter County in Pennsylvania contains poor deer range, and approximately 35 percent of legal bucks taken in this county by hunters are spikes. In Crawford and Venango Counties where range is considered adequate, only 4 percent of the bucks harvested fall into this category. In south-western Pennsylvania where deer are less plentiful and food is abundant, the bucks are heavier and carry larger antlers.

The herd size in Pennsylvania is now maintained at between 400,000 and 600,000 deer, yet the annual harvest is as great as during the peak population years around 1939. The combined harvest of bucks and does makes up about 20 to 25 percent of the total herd, and this kill is well within the surplus produced each year since the annual fawn crop is approximately 200,000.

Because the large predators, wolves and mountain lions, had long since disappeared from Pennsylvania, the surplus from the herd of 1,000,000 deer in the late 1930s and early 40s must have succumbed to malnutrition or starvation. There is little evidence they died of old age (maximum longevity of deer is at least twenty years), and much evidence to the contrary.

Inclement Weather

Severe weather conditions such as deep snow, severe cold spells, and drought often cause considerable mortality among wildlife, but such losses are seldom documented accurately. Also, the long-range effects of climatic factors are seldom considered. The consequences of one such climatic event, a sudden severe cold spell on the Texas coast during January 1940, on commercial catches of flounders was fully documented by Gordon Gunter.[12] The catches of flounders three months before and three months after the cold snap show that the percentage loss was the same regardless of locality and size of the original population (see Table 14). The population estimates were relative rather than absolute, being based on the assumption that fishing effort was the same in all localities and the same before and after the cold snap. The catches were presumed to reflect population size. These data illustrate the principle that climatic factors usually act independently of density.

Table 14. Three Month Catches of Flounders Before and After a Sudden Severe Cold Spell on the Texas Coast of the Gulf of Mexico[12]

Locality	Commercial Catch Before	After	Per Cent Decline
Mata Gorda	16,919	1,089	93.6
Aransas	55,224	2,552	95.4
Laguna Madre	2,016	149	92.6

Unfavorable weather conditions may combine with another environmental factor to induce mortality that would probably not have occurred if either agent had been operating singly. Snow depth of 20 inches is critical with deer, since it not only restricts movement but also places a high energy requirement on these animals. Deep snow prevents deer from reaching food. These conditions are lethal if food supplies are limited or if the deer are already weak from starvation. Approximately 80 percent of the losses in deep snow occur to the fawn age class; the next highest losses occur in the older females

(Pennsylvania data).

Disease

Disease is defined as illness; sickness; interruption or perversion of function of any of the organs; a morbid change in any of the tissues; or an abnormal state of the body as a whole. Disease may continue for a long or short period of time. The organism affected by disease may recover or the condition may terminate in death.

Traumatic injuries (disease) are caused by physical force or mechanical energy. Other physical agents capable of causing disease are changes in atmospheric pressure, sound waves, heat, cold, and electricity. Chemicals and radiation complete the list of inanimate agents capable of producing disease. Thus in a sense, the account of mortality induced by accidents, habitat destruction, predation, parasitism, malnutrition, starvation, and inclement weather (p.101) was a discussion of disease.

The infectious and parasitic diseases are caused by living organisms: rickettsias, viruses, bacteria, fungi, protozoans, and metazoans. Infectious and parasitic disease as a cause of mortality among wild animals is a matter of considerable interest, but relatively little is known of its natural history. That wild animals succumb to such diseases in large numbers has long been recognized, and for a time cyclic population declines were attributed to their occurrence. Originally, biologists thought that spread of infectious agents and parasites was facilitated by dense populations, hence population increases would inevitably be followed by epizootics of disease. But now it might be fair to say there is a glimmer of light at the end of the tunnel. Whether or not a pathogenic organism (or parasite) infects its victim depends upon a complex interaction between pathogenicity or virulence on the one side and the natural resistance of the host on the other. The results of the infection are likewise dependent on lethality of the infective agent and the mechanisms of resistance and defense of the infected animal or plant. The matter of contact between animals to spread infection may not be a very important element in the larger scheme of things.

Injurious or pathogenic agents cause death of cells, degenerative changes in cells and tissues, disturbance of metabolism, and interference with circulation of the blood and lymph. In some instances, disease is a result of congenital defects in cellular metabolism, general physiology, or anatomical structures of the organism. Thus in addition to the diseases mentioned there are congenital, auto-immune, allergic, neoplastic, and degenerative diseases.

Disease Caused by Chemicals

Chemical poisoning in nature was probably almost non-existent before the Industrial Revolution, but now its importance overshadows many other factors as a cause of death and population decline. A few illustrations will suffice to remind us of how really tragic the consequences of industrial pollution have been.

The mighty Atlantic salmon, a favorite of both commercial and sport fishermen, was once one of the most prevalent fishes in the Atlantic drainage areas. Its distribution extended from Kara in north-eastern Russia along the coast of Europe to Douro in the north-western part of the Iberian Peninsula. It spawned in the rivers of Iceland, Greenland, Canada, and the north-eastern United States. Salmon have almost completely disappeared from the Rhine, Wesser, and Elbe Rivers in Germany. A few isolated fishes still appear in the lower Rhine, but these are inedible because their fat contains phenol chemicals from the polluted waters. In Europe the decline of the salmon populations began in the mid-1890s as the rivers became polluted. In England industrialization started earlier, thus the Thames, once a favorite haunt of the salmon, lost its migrating salmon in 1830.

At the turn of the century, German fishermen repeatedly harvested more than 1,000 tons of salmon each year. Such catches cannot be made now. Today most European salmon are caught in Norway and Denmark. Canada leads the world with an annual catch amounting to 1,500-1,800 tons.

Insecticides

The story of chemical pesticides is only too well known to the majority. The best known of these chemicals is the chlorinated hydrocarbon DDT (dichloro-diphenyl-trichloroethane), which accumulates in the body fat of animals and is passed from prey to predator up the food chain and concentrated at each step until concentrations become lethal. In insects and other animals, DDT acts primarily on the central nervous system with effects ranging from hyperexcitability to death following convulsions and paralysis. Chronic effects on vertebrates include fatty infiltration of the heart and fatty degeneration of the liver which is often fatal. Fishes and other aquatic animals appear to be especially sensitive to DDT and other chlorinated hydrocarbons. Oxygen uptake is somehow blocked at the gills, causing death from suffocation. Chlorinated hydrocarbons are also known to slow the rate of photosynthesis in plants and marine phytoplankton.[13] Other chlorinated hydrocarbons designed to kill insects are benzenehexachloride (BHC), dieldrin, endrin, aldrin, chlordane, lindane, isodrin, and toxophene.

DDT also has its more subtle effects. As a result of the concentration of this chemical as it moves up the food chain, the danger to the life and reproductive capacity of fish-eating birds is extreme. Nesting failures attributable to DDT among American bald eagles *(Haliaetus leucocephalus)* and ospreys in parts of the United States have now reached proportions that bring the survival of these species into severe jeopardy. In addition, reproductive difficulties in populations of such diverse birds as the peregrine falcon, duck hawk, the brown pelican *(Pelecanus occidentalis),* and the Bermuda petrel *(Pterodroma cahow)* have been traced to residues of DDT and other chlorinated hydrocarbon insecticides. These chemicals interfere with calcium metabolism, making the eggshells so thin they are crushed by the weight of the incubating parents. Similar effects have been suggested for polychlorinated biphenyls (PCB). The PCBs may be five times more potent as killers than DDT. An attempt is now being made to assist endangered populations of bald eagles and ospreys by transplanting eggs from the Chesapeake Bay area where DDT is not a problem to the nests of the few remaining birds in affected areas. The success of these salvage operations will not be apparent for a long time.

The evidence against DDT in the case of its destructive influence on eggshells is now overwhelming. Eggshells of predatory birds in several museum collections have been measured. There is nearly always a sharp drop in eggshell thickness for the period 1945-1947, when DDT was generally introduced. In Britain a strong correlation has been shown between the level of insecticide contamination in various geographical regions and eggshell thickness in peregrine falcons, sparrow hawks *(Accipiter nisus),* and golden eagles *(Aquila chrysaetos)* nesting in those regions.

Chlorinated hydrocarbons have great stability. It is not entirely clear how long DDT persists in ecosystems. Fifty percent of the DDT sprayed in a single treatment may still be found in a field ten years later. This does not mean, however, that the other 50 per cent has been degraded to biologically inactive chemical molecules; it may only have gone somewhere else. Probably DDT, including its biologically active breakdown product DDE, has an average half-life (time required before 50 percent has been degraded) of much more than ten years. Indeed, some authorities suggest DDE may be virtually indestructible.

DDT and other chlorinated hydrocarbons are now widespread in nature.[14] Up to 32 pounds per acre of DDT were found in the upper layer of mud in the Long Island Estuary. Such concentrations in the soils of the United States are not unusual. Concentrations of DDT in the fat deposits of Americans average 7-12 ppm (parts per million), and the people of India and Israel have much higher concentrations. More startling and significant in some ways has been the discovery of DDT residues in the fat deposits of Eskimos and in Antarctic penguins and seals. Seals from the east coast of Scotland have been found to have concentrations of DDT as high as 23 ppm. in their blubber. The only saving grace is the fact that the chemical is stored in adipose tissues far removed from the sensitive nervous system and thus rendered relatively harmless.

Oil Pollution

Oil as a major ocean pollutant has received a great deal of publicity, mainly because much of it is the result of spectacular accidents. The Torrey Canyon tanker disaster near the English coast, the two-tanker collision in San Francisco Bay in 1971, and the oil tanker-chemical ship collision in Delaware Bay in 1975 are the best-known examples. Each of these oil spills resulted in the deaths of thousands of water-fowl and shore birds. The oil fouls the bird's feathers and destroys their capacity to repel water and insulate the body from the cold. The birds ingest the toxic oil while attempting to preen their feathers. Unable to dive for food because of oil-soaked feathers, many of the birds starve or succumb to cold exposure. Besides killing directly, oil toxins may also reduce egg viability. Surviving birds may not reproduce at all the year following the contamination.

Attempts to rescue oil-soaked waterfowl are well-meaning but doomed to failure from the start. I conducted postmortem studies on a few of the birds killed in San Francisco Bay and followed nearly 100 rehabilitated scoters from an oil spill in Delaware Bay in 1966. The birds are severely stressed by the experience of being captured and cleaned with soaps and detergents. Some of the Delaware Bay birds were even cleaned with high frequency sound waves. Mortality within the first few days of a spill is usually close to 50 percent. The scoters in question suffered severe mortality seven months after the rescue operation when they were captured and handled for leg-banding. Less than five percent of the 100 scoters were alive in captivity one year later. The systemic stress of the initial oil soaking (starvation and cold exposure) combined with the stress of capture, handling, and captivity is usually too much for such vulnerable wild creatures to withstand.

These massive accidental spills from oil tankers probably account for less than 0.1 percent of the total oil transported at sea, but unfortunately the total volume transported is so large, about 360 billion gallons per year, that the spills amount to a considerable quantity. Oil also reaches the sea in connection with refueling operations and is discharged from the pumping of bilges. Contaminated refined petroleum is also dumped at sea, because it is cheaper or more convenient than bringing it ashore for treatment. In addition to shipping spills, there are accidents in extraction from sea-floor drilling, of which the Santa Barbara leak off the coast of California is the best known example. Finally, some oil reaches the sea in sewage wastes, and natural oil pollution occurs from undersea faults and joints.[14]

Given time and protection, increased reproduction and decreased mortality in depleted water-fowl populations may replace the birds lost to oil pollution. To the jackass penguins *(Spheniscus demersus)* already reduced to less than four percent of their original population, oil pollution may be the final insult. Dassen Island, 32 miles from Cape Town, South Africa, is a major refuge for these 15-inch-tall birds. In 1930, there were 5,000,000 of these penguins on the island. But now on Dassen Island and on nearby islands at the tip of Africa, up its west coast and along the Indian Ocean off Port Elizabeth there are only 60,000 to 200,000 birds remaining.[15]

Jackass penguin populations were exploited because their eggs were treasured for a special flavor and the somewhat greenish white that turns to translucent jelly when cooked. Eggs were worth as much as $2.00 a dozen at mainland markets. Egg raids were prohibited by the South African government in the late 1960s.

The penguins now face a more serious menance. With the closing of the Suez Canal, 650 oil tankers a month were rerouted around the southern coast of Africa. Consequently, spills and accidents soaked many of the penguins in oil, killing them by the tens of thousands. Even a reopened Suez cannot solve the problem, for the supertankers of today are too large to fit through it.[15]

Most of our knowledge of the potential effects of oil pollution on oceanic ecosystems comes from studies of a relatively small oil spill that occurred in 1969 near the Wood's Hole Oceanographic Institute in Massachusetts.[16] The immediate and obvious destruction of populations of fish, shellfish, and sea birds at the sites of oil spills has been well publicized. The long-range effects, however, vary with the type of oil, the distance from the shore where it spilled, how long it can be weathered (degraded by microorganisms, dissolve and evaporate) before reaching shore, and what organisms live there. Some components of oil are toxic, while others are chemicals capable of inducing cancer; weathering can reduce their toxicity, but the carcinogenic components are long-lasting. Oil washed to shore lingers on rocks and sands for months or years, but the marine life may need a decade or more to recover, even after the oil is no longer obvious.

Detergents used to clean up oil spills have been found to worsen the situation in many cases. Not only are the detergents themselves toxic to many forms of life, but they disperse the oil and spread it into new areas. By breaking up the oil into droplets, they may also render it more easily absorbable by small marine organisms.

Immediately after the Massachusetts oil spill, there was a 95 percent mortality of fish, shellfish, worms, and other marine animals. Repopulation had still not taken place nine months later. Surviving mussels failed to reproduce. Some constituents of the oil were still killing bottom-dwelling organisms eight months later. Surviving shellfish and oysters took in enough oil to be inedible and retained it even months after distant transplanting. According to marine biologist Max Blumer, there is a possibility that the carcinogenic compounds in oil, absorbed and unchanged in phytoplankton and other small organisms, may in time be incorporated into and contaminate entire food chains.

No one knows how long we can continue to pollute the seas with insecticides, polychlorinated biphenyls, oil, mercury, cadmium, and thousands of other pollutants without bringing on a collapse of ocean productivity. Subtle changes may already have started a chain reaction in that direction, as shown by declines in many fisheries, especially those in areas of heavy pollution caused by dumping of wastes.[14]

Diseases in Natural Populations

In general, the overall causes of death in natural populations have not been investigated in any comprehensive manner. Scattered reports in the literature indicate a complex array of disease-producing agents in natural environments. As one ecologist explained, the actual cause of death is relatively unimportant in population problems. It is always complex and can rarely be assigned without doubt to a particular item. There is probably some truth to this statement, but it should not be misconstrued. It does not mean that DDT, mercury, and oil in the environment are not killers. It means only that in natural, undisturbed ecosystems the causes of death are not so important as the overall population dynamics and long-term population effects. For example, a moose weakened by hydatid cysts in the lungs is caught and killed by wolves. Was the cause of death parasitism or predation? What was the carrying capacity of the moose's home range?

I was able to compare patterns of disease in a captive population of woodchucks with those occurring in a natural population of the same species. I collected and autopsied 1,007 woodchucks from a 10,000-acre study area in south-central Pennsylvania between 1957 and 1959. I trapped 100 young woodchucks from the same study area between 1960 and 1975 and kept them in cages at the Penrose Research Laboratory. The 100 woodchucks were trapped as young of the year during the summer when they were three to four months of age. Figure 12 is a graphic illustration of the survival curves for wild and captive woodchucks. The mean age at death of 76 captive woodchucks was 53 months compared with 12 months for the woodchucks living in the wild. Incidentally such differences between survival rates of captive and natural populations of a species are characteristic.

Figure 12. Survival Curves for Wild and Captive Woodchucks

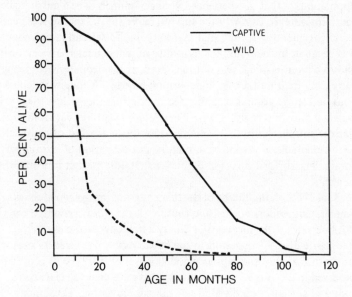

The survival curve for wild woodchucks was calculated from the data in Table 15. Woodchucks were captured and marked individually with numbered metal ear-tags. Young of the year were captured when they

Table 15. Comparison of the Percentage of Young Woodchucks
Recovered with that of the Older Woodchucks Recovered.

	Young	Old
Released one year	369	287
Recovered the next	103	147
Per cent recovered	27.9	51.2

were two to six months of age. The 'old' woodchucks were at least 11
months of age when marked and released. Recovery by trapping the
following year was considered evidence of survival for one year.
Woodchucks that moved out of the study area or avoided retrapping
were also listed as lost, however the error was considered minimal since
the mortality calculated from trapping data agreed with estimates of
population size and population turnover in the study area. The rate of
disappearance of younger woodchucks was nearly twice that of the
older woodchucks, illustrating again a consistent discovery in natural
populations — that inexperienced younger animals have a much higher
mortality rate. In captivity the survival curve for the species was nearly
a straight line, thus differential mortality due to age and inexperience
was negated. Incidentally, the mortality of woodchucks between birth
and two months of age was not measured, as the neonates were being
nursed by their mothers in underground burrows. Young woodchucks
appeared above ground accompanied by their mothers when they
weighed only 300 to 400 grams. They seldom ventured far from the
burrow entrance until they were two months of age, but of course they
were vulnerable to predation during the critical period of learning to
forage. The survival curves in Figure 12 therefore start at four months
of age.

Table 16 lists the causes of death among the captive woodchucks.
Cancer, glomerulonephritis, and rupture of the aorta accounted for
47.3 percent of the mortality. Primary carcinoma of the liver,
neoplastic disease originating in the hepatocytes, was listed as a cause
of death for 14 of the captive woodchucks, but this type of cancer was
observed in 22 of the 76 woodchucks that have died thus far. Also,
ruptured aorta was observed in ten animals altogether. Thus, many
captive animals were afflicted by more than one disease, although only
one is listed as the primary cause of death. Glomerulonephritis, a
disease involving thickening and degenerative changes in the basement

Table 16. Causes of Death Among Woodchucks Confined in the
Laboratory

Age is measured in months

Cause of Death	Number	Mean Age at Death	Range in Ages	Per Cent Mortality
Cancer of Liver*	14	54	23 — 106	18.4
Glomerulonephritis*	11	58	12 — 101	14.5
Ruptured Aorta*	8	34	9 — 74	10.5
Bacterial Infections*	6	70	28 — 97	7.9
Cerebral Hemorrhage	5	38	28 — 54	6.6
Myocardial Infarction	4	61	46 — 93	5.3
Larval Tapeworm Inf.*	4	54	30 — 72	5.3
Bronchopneumonia*	4	57	5 — 105	5.3
Hepatitis*	3	59	45 — 66	3.9
Undetermined	3	51	21 — 90	3.9
Other Neoplasms	3	71	62 — 79	3.9
Congestive Heart Failure	2	68	64 — 73	2.6
Viral Infection	2	10	5 — 15	2.6
Miscellaneous Diseases	7	54	6 — 86	9.3
Totals:	76	53	5 — 106	100.0

Indicates same disease noted among population of wild woodchucks.

membranes of the glomerular capillaries, is generally considered an
auto-immune disease of uncertain etiology.

The fact that many of the diseases observed in captive woodchucks
were also observed in wild woodchucks is noteworthy. Most of the wild
woodchucks were either trapped or killed by hunters in the study area,
hence most of them were healthy or demonstrated only early stages of
disease when examined postmortem. Wild woodchucks afflicted by
serious illnesses or in terminal stages of disease would probably remain
underground, thus the chances of recovering such animals would be
remote. Thus, the one example of aortic disease in a wild woodchuck
was represented by an older specimen with a calcified aortic arch.
While all of the captive animals with ruptured aortas died within

approximately 24 hours of the episode, the healed lesion in the wild woodchuck would suggest that this disease is not always lethal.

Cerebral hemorrhage (stroke) and myocardial infarction (heart attack) were not observed in wild woodchucks, although arteriosclerosis (hardening of the arteries), the disease responsible for stroke and heart attack, existed among many of them. Arteriosclerosis is a degenerative disease that takes time to progress to the stage where it interferes with circulation of the blood. Thus, the disease would not be expected to have much influence on the mortality rate of a natural population where rapid annual turnover precludes the accumulation of aged animals.

The tapeworm larvae infecting the captive woodchucks must have been picked up while these animals were living in their native habitat. The captive woodchucks were fed the standard Philadelphia Zoo diet, which has not been a source of tapeworm infection for the Zoo animals.

The wild woodchucks showed a high incidence of bronchopneumonia during the late winter and early spring following the hibernation period. Some of these infected animals must have died, but there was no way to ascertain the magnitude of such losses. Pulmonary lesions were observed in males coming above ground in February, thus it would appear that the physiological state existing during hibernation may have been conducive to the development of bacterial infections of the lungs.

In general, a comparison of disease patterns in captive and wild animals would suggest a different set of environmental influences at work. Wild animals are constantly exposed to predators, hunters, vehicular traffic, adverse weather conditions, variable food supplies, parasites, and various pathogenic agents. Under these circumstances, they infrequently reach the age at which degenerative and neoplastic diseases would begin to take their toll. For example, neoplastic disease, a process of uncontrolled growth of individual cells, is caused by a number of physical and biological agents (radiation, chemicals, viruses, etc.) The incidence of cancer among the birds and mammals of the Philadelphia Zoo is only 2.0 percent during the first quarter of life but increases to 40.7 percent in the last quarter. The chances of finding a high incidence of cancer among natural populations would be unlikely under the circumstances.

Rapid Turnover in Nature

The average person assumes that mortality among natural populations is similar to that experienced by man. Such persons are understandably

amazed when confronted with statistics derived from field studies. An
enthusiastic experimental pathologist once came to me for advice
before embarking on a study of arteriosclerosis in the wild cottontail
rabbit. His reasoning was partly sound. The domestic rabbit has long
been used in experiments designed to determine how cholesterol feeding
induces hardening of the arteries. He thought it would be profitable to
examine wild rabbits to determine the incidence of the disease under
natural conditions. As mentioned before, this degenerative disease takes
time to develop, thus it would be necessary to examine a number of
rabbits at least four years old. A significant sample size would be in the
order of 40 or 50 animals. The results of a study conducted on the
Conemaugh River Reservoir in western Pennsylvania dampened the
pathologist's spirits immeasurably. These data are graphed as a
survivorship curve in Figure 13. This study on 500 acres required 241
days of trapping and 33,931 total trap nights (one trap set overnight
equals one trap night) to catch and mark 924 rabbits. The survivorship
curve shows a very high mortality for juvenile cottontails. Fifty percent
of the rabbits trapped survived less than four months. About ten
percent survived one year. Only one animal was known to live as long
as three years on the study area.

Figure 13. Survivorship Curve for Cottontail Rabbits

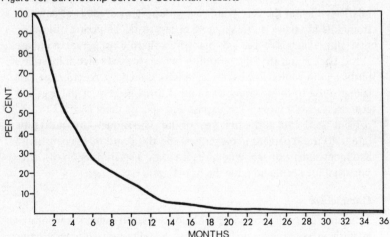

The experimentalist made a few rapid calculations and gave up his
promising idea. If he had lowered his sights slightly and settled on
three-year-old rabbits as suitable subjects for examination, he could have
collected 40 three-year-old animals with an expenditure of 1,357,240

trap nights of effort. He would also have had to expand the study area to 20,000 acres.

The utilization of trapping data to construct a survivorship curve does require certain assumptions. In this case one must assume that cottontails once trapped do not become 'trap shy.' We shall see subsequently that this is not always a valid assumption, but the conclusions about high mortality were warranted in the Conemaugh River study area. The area was open to public hunting which afforded research biologists the opportunity to recover animals ear-tagged during the trapping operations.

In a period of ten years during which field studies were conducted on the Conemaugh River Reservoir, hunters shot between 20 and 25 percent of the rabbits trapped and marked before the hunting seasons. Hunters also killed a number of untagged rabbits which afforded a measure of the trapping efficiency and a double check on the survival data. Again only 10 percent of the tagged rabbits in the hunters' bags were more than one year old. Similar studies of cottontail rabbits in a different part of the United States found the same rapid turnover. The probability of dying during the first four months of life was 0.74.

Mortality and Age

It is generally recognized that the total death rate from all causes is relatively high in the very young age groups in most kinds of animals. In middle life there is often a more or less sustained period of low mortality which is succeeded finally by a sharply rising rate in old age. David Lack states that the mortality rate is always higher in juvenile birds than in adults. For example, in passerines 45 percent of the eggs laid give rise to flying young and only 8 to 18 percent of the eggs laid give rise to adult birds. Thus in passerine species there is an 82 to 92 percent death rate in the first year of life. The annual adult mortality rate is 40 to 60 percent in passerines, and this death rate is constant and independent of age. Wild birds live only a small fraction of their potential life span, and none die of old age.

Cannibalism

Litter mortality in confined populations of small rodents is wholly intrinsic, depending on such factors as the failure of milk-secretion in the mother, desertion, destruction of the nest during aggressive fighting among the adults and cannibalism. Sometimes, and perhaps generally, it involves the social status of the mother, those lower in the scale incurring much heavier losses than the dominant females.[17]

Cannibalism, especially among neonatal young, is extremely common in laboratory and domestic animals. It is also a common cause of mortality in zoological gardens.

Precise details about cannibalistic behaviour are difficult enough to obtain from captive animals. In the wild they are acquired infrequently because of the extreme secretiveness of females with young. Close confinement of caged animals with undue noises and distractions in close proximity tends to aggravate postnatal losses from cannibalism, desertion, and fighting among adults. There are many scattered references describing first-hand observations of infantile mortality induced by wild parents. Wynne-Edwards[18] believes that at least some post-fledgling mortality in various species of birds is intrinsic, promoted by social conflict, and therefore an adaptive mechanism for population control.

Almost every litter of lions contains one or more cubs weaker than the rest. When food is in short supply infant mortality is high because the females have insufficient quantities of milk.[18] Losses continue after weaning during the long period that the cubs accompany their mothers. The same authority is quoted as saying that a great number of young lions between nine months and two years of age fall victims of their own kind.

Breder and Coates[19] founded two colonies of guppies *(Lebistes reticulatus)* in identical aquariums, the first with a single gravid female and the second with a mixed group of 50 fish. Food was supplied in excess. In the course of about six months, both populations reached the same asymtote, namely nine fish. In the first tank, survival of the initial brood of young was 100 percent. In succeeding broods, it was progressively less until when the fourth brood appeared, all young were eaten by the mother. By this time the biomass of her earlier broods had increased considerably, and she began to thin the fish from these broods also. In the second tank, no young were ever allowed to survive, however briefly after birth. All were immediately eaten by the group of fish already present in the tank. Moreover, the latter consumed one another by degrees until here again the final equilibrium population of nine was reached. Infanticide in Lebistes was directly proportional to population concentration.

Longevity

It seems appropriate to discuss longevity in terms of the two types recognized by population students.[20] Physiological longevity represents the capacities of the individuals of a species to live out their life spans, while ecological longevity depicts the observed life duration of the members of a natural population. Physiological longevity is also defined as the maximum life span attained by each species-population when living under optimum ecological conditions and having a genotype homogeneous for all components.[20]

Life span can be regarded as a physical dimension of a species on the same footing as the linear or mass dimensions. Larger animals generally have longer life spans than smaller ones. Further, animals that mature slowly over a number of years have longer life spans than those that mature rapidly. Age at sexual maturity correlates generally with life span also, and animals that require several years to reach sexual maturity have correspondingly long life spans and low reproductive rates.

The maximum exhibition periods of mammals living in the Philadelphia Zoo listed in Table 17 provide some idea of the potential maximum life spans of various mammalian species. The data were compiled by the taxonomic grouping of families which admittedly obscures variability between genera and between species within the family. The genotype was far from homogeneous for all components and also was not likely to have been 'healthy' for all components. Finally, ecological conditions in the Zoo were not optimal in spite of every effort to make them so.

Physiological longevity and life span are academic terms that will never be known with certainty for any species. But maximum exhibition periods attained by mammals living in the environment of the Philadelphia Zoological Garden do illustrate clearly that some species characteristically live much longer in captivity than others. Analysis of these data does not give the exact physical dimension of life span for a family group, but it does indicate the relative magnitude of differences between life spans of family groups.

Exhibition periods are not true life spans since many animals come to the Zoo as wild-trapped specimens. Thus, the life span in most cases would be somewhat longer than the exhibition period. The word 'potential' should be emphasized again, because the figures in Table 17 will change as new data are collected. At the moment, two orangutans *(Pongo pygmaeus)* living in the Philadelphia Zoo are 55 years old and the famous gorilla Massa *(Gorilla gorilla)* is at least 42 years old and still alive and well.

Life Tables

The life table is a well-known method of presenting data on longevity and age-specific mortality. The standard life table is constructed from

Table 17. Members of Each Family of Mammals that Lived Longest in Captivity at the Philadelphia Zoo[21]

	Common Name	Scientific Name	Sex	Months
Order: Artiodactyla				
Bovidae	African buffalo	Syncerus caffer	♀	315
Camelidae	Dromedary camel	Camelus deromedarius	♂	341
Cervidae	Malayan sambar deer	Cervus unicolor	♂	233
Hippopotamidae	Common hippo-potamus	Hippopotamus amphibus	♂	433
Suidae	Wild boar	Sus scrofa	♂	194
Tayassuidae	Collared peccary	Tayassu tajacu	♀	203
Order: Carnivora				
Canidae	Red wolf	Canis niger	♀	177
Felidae	Leopard	Panthera pardus	♂	233
Hyaenidae	Spotted hyena	Crocuta crocuta	♂	298
Mustelidae	American badger	Taxidea taxus	♂	242
Procyonidae	White-nosed coati	Nasua nasua narica	♀	177
Ursidae	Japanese brown bear	Ursus arctos lasiotus	♀	387
Viverridae	Binturong	Arctictis binturong	♀	216
Order: Edentata				
Bradypopidae	Two-toed sloth	Choloepus didactylus	♀	278
Dasypodidae	Nine-banded armadillo	Dasypus novemcinctus	♀	124
Myrmecophagidae	Giant anteater	Myrmecophaga tridactyla	♀	62
Order: Hyracoidea				
Procavidae	Cape hyrax (2 specimens)	Procavia capensis	2♀	106
Order: Insectivora				
Erinaceidae	European hedgehog	Erinaceus europaeus	♀	50
Order: Marsupialia				
Didelphidae	Philander opossum	Caluromys philander	♀	25
Macropodidae	Great grey kangaroo	Macropus canguru	♀	198
Order: Monotremata				
Tachyglossidae	Australian short-beaked echidna	Tachyglossus aculeatus	♀	593
Order: Perissodactyla				
Equidae	Mongolian wild horse	Equus przewalskii	♂	364
Tapiridae	Brazilian tapir	Tapirus terrestris	♂	195
Rhinocerotidae	Indian rhinoceros	Rhinoceros unicornis	♀	239

Table 17 (Continued)

	Common Name	Scientific Name	Sex	Months
Order: Pinnepedia				
Otariidae	California sea lion	Zalophus californianus	♀	246
Order: Primates				
Tupaiidae	Common tree-shrew	Tupaia glis	♀	26
Callithricidae	Common marmoset	Callithrix jacchus	♂	74
Cebidae	Tufted capuchin	Cebus apella	♀	247
Cercopithecidae	Moor macaque	Macaca maurus	♂	337
Lemuridae	Mongoose lemur	Lemur m. mongoz	♀	304
Lorisidae	Slow lorris	Nycticebus coucang	♂	99
Pongidae	Chimpanzee	Pan troglodytes	♂	454
Order: Proboscidea				
Elephantidae	Indian elephant	Elephas maximus	♀	457
Order: Rodentia				
Castoridae	Canadian beaver	Castor canadensis	♂	125
Caviidae	Patagonian cavy	Dolichotis patagonum	♂	125
Cricetidae	Greater Egyptian gerbil	Gerbillus pyramidum	♂	84
Dasyproctidae	Golden agouti	Dasyprocta agouti	♂	186
Erethizontidae	Canadian porcupine	Erethizon dorsatum	♀	94
Heteromyidae	Richardson's kangaroo rat	Dipodomys ordii	♂	118
Hystricidae	Brush-tailed porcupine	Atherurus africanus	♀	275
Sciuridae	Western fox squirrel	Sciurus niger	♀	119
Order: Tubilidentata				
Orycteropodidae	Cape aardvark	Orycteropus afer	♂	43

a 'cohort' of animals all born at the same time. Table 18 is such a table constructed from data on cottontail rabbits. Such tables are obviously difficult to prepare for wildlife studies.

Life tables can also be constructed if the age structure of a population is known. This can be accomplished if a method of determining age is available. The sequence of tooth eruption and patterns of wear are valid criteria of age for both mule deer and white-tail deer. The age class distribution of 661 adult male white-tails killed by hunters in Pennsylvania in 1954 is contained in Table 19. The probability of dying during one year for antlered deer can be calculated by dividing the

Table 18. Life Table for Cottontail Rabbits[22]

Age Interval in Months	Number Alive at Beginning of Month of Age	Number Dying During Month of Age	Mortality Rate (number dying per 1000 alive at beginning of month)	Complete Expectation of Life (average number of months of life remaining at beginning of month of age)
X	1_x	d_x	$1000q_x$	e_x
0 to 4	10,000	7,440	744	6.5
4 to 5	2,560	282	110	6.6
5 to 6	2,278	228	100	6.5
6 to 7	2,050	246	120	6.5
7 to 8	1,804	307	170	6.4
8 to 9	1,497	150	100	6.4
9 to 10	1,347	175	130	6.3
10 to 11	1,172	164	140	6.3
11 to 12	1,008	212	210	6.3
12 to 13	796	143	180	6.3
13 to 14	653	98	150	6.2
14 to 15	555	55	100	6.0
15 to 16	500	65	130	5.8
16 to 17	435	31	70	5.6
17 to 18	404	24	60	5.3
18 to 19	380	49	130	5.0
19 to 20	331	36	110	4.9
20 to 21	295	47	160	4.6
21 to 22	248	20	80	4.4
22 to 23	228	39	170	4.2
23 to 24	189	32	170	4.0
24 to 25	157	13	80	3.7
25 to 26	144	7	50	3.4
26 to 27	137	30	220	3.1
27 to 28	107	12	110	2.9
28 to 29	95	13	140	2.6
29 to 30	82	32	390	2.4
30 to 31	50	7	140	2.3
31 to 32	43	9	210	2.1
32 to 33	34	11	330	1.9
33 to 34	23	16	700	1.9
34 to 35	7	3	350	2.3
35 to 36	4		...	2.0
36 to 37	4		...	1.5
37 to 38	4		...	1.0
38 to 39	4	4	1,000	0.5

Table 19.   Life Table Based on Age Class Distribution of 661 Adult Male White-tail Deer Killed by Hunters in Pennsylvania in 1954

X (years)	1_x	d_x	1000q_x	e_x (years)
½ — 1½	661	285	431	1.46
— 2½	376	202	537	1.19
— 3½	174	120	690	1.00
— 4½	54	36	667	1.11
— 5½	18	8	444	1.33
— 6½	10	7	700	1.00
— 7½	3	1	333	1.16
— 8½	2	2	1,000	0.50

number of animals in the first age class by the total number of animals. In this case 285 ÷ 661 = 0.431. The age structure reflects the average annual mortality if the recruitment rate to the population is relatively constant, which would be the case in a stable population. If the age structure of the population is determined every year, the error due to variations in recruitment can be averaged out.

A more exact life table can be constructed for captive populations since all deaths can be enumerated, and age or length of captivity can be assigned to each dead animal. Table 20 is a life table constructed for mammals dying in the Philadelphia Zoo between 1901 and 1964. Some investigators have guessed that wild animals would live longer in captivity because of the protection afforded them from starvation and predators. However, the data in Table 20 suggests that age-specific mortality patterns of captive and natural populations of wild mammals are not necessarily so dissimilar. The mortality rates among young Zoo mammals, those on exhibition for the first twelve months, were quite high. The family Didelphidae, for example, had losses of 96 percent during the first twelve months on exhibition between 1901 and 1934. Annual mortality rates decreased considerably after 1935 because of improvements in nutrition and husbandry, but losses were still relatively high during the critical first year of captivity. Thus mortality data from zoo mammals are fairly representative of the general trend in mortality patterns. The young incur the highest losses, and few mammals come close to living out their physiological life spans. The four quarters of life are based on maximum exhibition periods recorded for the Philadelphia Zoo.

Table 20. Life Tables Showing Proportion of Mammals That Died During Each Quarter of Life (20 families) in the Philadelphia Zoo. Both Sexes are Included in the Tables[21]

| | 1901–34 | | | | | | 1935–64 | | | | | |
---	N	<12	I	II	III	IV	N	<12	I	II	III	IV
Order: Artiodactyla												
Bovidae	283	.34	.36	.18	.10	.02	157	.33	.31	.23	.10	.03
Camelidae	43	.19	.41	.26	.02	.02	24	.21	.38	.21	.03	.17
Cervidae*	288	.34	.37	.20	.08	.01	162	.30	.30	.23	.12	.05
Order: Carnivora												
Canidae*	187	.38	.32	.19	.08	.03	124	.30	.22	.23	.19	.06
Felidae*	182	.46	.26	.18	.08	.02	112	.30	.28	.21	.12	.09
Mustelidae*	129	.50	.40	.07	.02	.01	72	.18	.60	.08	.08	.06
Procyonidae*	126	.38	.40	.15	.05	.02	67	.15	.37	.33	.12	.03
Ursidae*	63	.21	.49	.21	.08	.01	32	.16	.16	.31	.09	.28
Viverridae	58	.21	.53	.14	.10	.02	24	.09	.33	.25	.21	.12
Order: Marsupialia												
Didelphiidae*	218	.96	.02	.01	.00	.01	27	.59	.30	.04	.07	.00
Macropodidae*	115	.42	.42	.11	.04	.01	70	.46	.17	.18	.10	.09
Order: Pinnepedia												
Otariidae	34	.44	.32	.18	.03	.03	22	.23	.23	.27	.09	.18
Order: Primates												
Callithricidae	71	.70	.23	.03	.02	.02	43	.72	.09	.03	.09	.07
Cebidae*	207	.50	.49	.01	.00	.00	89	.36	.42	.17	.03	.02
Cercopithecidae*	497	.43	.48	.06	.02	.01	164	.36	.38	.16	.07	.03
Lemuridae*	92	.48	.49	.03	.00	.00	9	.22	.34	.11	.22	.11
Pongidae	28	.39	.57	.04	.00	.00	46	.46	.33	.09	.03	.09
Order: Rodentia												
Castoridae	20	.70	.15	.05	.00	.10	17	.47	.12	.18	.18	.05
Erethizontidae	69	.71	.22	.04	.00	.03	41	.63	.24	.03	.10	.00
Sciuridae*	135	.58	.35	.05	.01	.01	57	.30	.38	.23	.05	.04

N = Total number of animals. <12 = Less than 12 months on exhibition. I–IV = Quarters of life based on the maximum potential life spans in Table 17. * = Mortality patterns differ significantly.

Formulas

As is true for birth rate, the demographers use both a crude death rate and various specific death rates. The crude death rate is:

$$R_c = \frac{D}{P}$$

where:

R_c = crude death rate

D = deaths

P = the total population or $D + (P-D)$.

This rate is usually expressed per 1,000, per 10,000 or per 100,000 individuals.

The specific death rate is:

$$R_s = \frac{D_c}{E}$$

where:

R_s = specific death rate

D_c = deaths in a specified class of the population

E = number exposed to risk of dying.

Instantaneous death rate (d) is the number that died during a specified period of time divided by the average population. The formula for the instantaneous death rate is:

$$d = \frac{D}{N_o + N_t/2}$$

where:

d = number of deaths during the specified time interval

N_o = Population at beginning of time interval

N_t = Population at the end of the time interval.

The average population during the year can also be calculated by the following formula:

$$P = N_1 + N_2 + \ldots N_{12}/12$$

where:

P = average population

N_1 = population at the end of the first month

N_2 = population at the end of the second month, etc.

The population can be calculated quarterly, semi-annually, or at whatever intervals are convenient, but accuracy depends on the number of population censuses during the year.

The probability of dying (q) is the number that died during a time interval divided by the number alive at the beginning of the interval. This probability is usually expressed as a percentage. Thus:

$$q = 100 - p$$

where 100 animals are alive at the beginning of the time interval and p is the number of survivors at the end of the time interval.

The life table is a summary of age specific mortality rates operating on a population:

$$l_x + 1 = l_x - d_x$$

$$q_x = \frac{d_x}{l_x}$$

The calculation of expectation of further life (e_x) is more tedious than complicated. First, obtain the average number of animals alive in each age interval, which is called the life table age structure (L_x):

L_x = number of animals alive on the average during the interval x.

$$L_x = \frac{l_x + l_x + 1}{2}$$

Then sum these cumulatively from the bottom of the life table and obtain a set of values expressed as animals x time units, which is called T_x:

$$T_x = \Sigma L_x$$

$$e_x = \frac{T_x}{l_x}$$

Summary and Conclusions

Mortality is the force for population decline and antithetic to reproduction and immigration. The cause of death is a starting point for theoretical treatments of mortality patterns within a population. Several questions are posed after the cause of death is determined. Etiology (the life history of disease) may be the first and most important question. What are the reasons for a particular animal dying from a particular causative agent? Insecticide poisoning may be the actual cause of death but etiology is also concerned with how the poisonous chemical got into the tissues, its lethal concentrations, its mechanisms of death, etc.

A second important question concerns the effects mortality has on long-term population size. Will the loss be recompensed with subsequent reproduction? If the loss is local, will its impact be minimized by dispersion (rearrangement of spacing patterns) or emigration?

A third question arises – does the mortality substitute for another mortality factor? Predation may be the immediate cause of death, but perhaps the prey was exposed because of population levels above the carrying capacity of the habitat or forced out of its normal home range by flooding. An interesting component of substitute mortality is related to infectious disease, longevity, and time. Mortality may be deferred until later in life under certain circumstances. People who once died of infectious diseases before the advent of effective drug therapy now live long enough to die of heart attacks, strokes, and cancer. The time factor can best be illustrated by hunting season statistics. Game animals killed by hunters in the autumn would have been killed by predators later or would have died because of adverse environmental conditions during the wintertime.

Finally, the extent of mortality during the year should be considered relative to the population's reproductive rate and the habitat's carrying capacity. A mythical animal population will serve as an example. Consider one that consists of one male and one female at time t ($N_0 = 2$) with an annual reproductive rate of two. In this situation the mortality rate (d) during the year must be 50 per 100 animals to maintain equilibrium. A rate below 0.50 will add animals to the population. Subsequently, if reproductive rate is not density-dependent, the population will grow. If the population is existing below the carrying capacity, the surviving offspring can be assimilated, but if carrying capacity is exceeded, a number of compensatory events will occur. All of this brings us to an important, inescapable conclusion. No population can continue to grow indefinitely.

Notes

1. C.B. Huffaker and C.E. Kennett, 1956. 'Experimental studies on predation: predation and cyclamen − mite populations on strawberries in California'. *Hilgardia* 26: pp.191-222.
2. G. Bump, R.W. Darrow, F.C. Edminster and W.F. Crissey, 1947. *The Ruffed Grouse, Life History, Propagation, Management.* The Holling Press, Inc., Buffalo, New York.
3. W.F. Crissey, and R.W. Darrow, 1949. *A study of predator control on Valcour Island.* New York State Conserv. Dept., Div. of Fish and Game, Res. Ser. No. 1.
4. F.C. Edminster, 1939. 'The effect of predator control on ruffed grouse populations in New York'. *J. Wildl. Mgt.* 3: pp.345-52
5. L.D. Mech, 1966. *The Wolves of Isle Royale.* Fauna of the National Parks of the United States, Fauna Series 7, US Government Printing Office, Washington, DC.
6. A. Murie, 1944. *The Wolves of Mount McKinley.* US Nat. Park Serv., Fauna Ser. 5, US Government Printing Office, Washington, DC.
7. I.M. Cowan, 1947. 'The timber wolf in the Rocky Mountain national parks of Canada'. *Can. J. Res.* 25: pp.139-74.
8. O.P. Pearson, 1966. 'The prey of carnivores during one cycle of mouse abundance'. *J. Anim. Ecology* 35: pp.217-33.
9. M.H. Stenlund, 1955. 'A field study of the timber wolf *(Canis lupus)* on the Superior National Forest, Minnesota'. *Minn. Dept. Cons. Tech. Bull.* 4.
10. G.C. Varley, 1947. 'The natural control of population balance in the knapweed gall-fly *(Urophora jaceana)'. J. Anim. Ecol.* 16: pp.139-87.
11. W.A. Riley, 1939. 'The need for data relative to the occurrence of hydatids and of *Echinococcus granulosa* in wildlife'. *J. Wildl. Mgt.* 3: pp.255-7.
12. G. Gunter, 1941. 'Death of fishes due to cold on the Texas coast, January 1940'. *Ecol.* 22: pp.203-8.
13. C.F. Wurster, 1968. 'DDT reduces photosynthesis by marine phytoplankton'. *Science* 158: pp.1474-5.
14. P.R. Ehrlich, A.H. Ehrlich and J.P. Holdren, 1973. *Human Ecology, Problems and Solutions.* W.H. Freeman and Company, San Francisco.
15. L. Milne and M. Milne, 1975. 'The last survivors'. *National Wildl.* 5: pp.13-14.
16. G. Coan, 1971. 'Oil pollution'. *Sierra Club Bulletin,* March issue, pp.13-16.
17. J.B. Calhoun, 1950. 'The study of wild animals under controlled conditions'. *Ann. N.Y. Acad. Sci.* 51: pp.1113-22.
18. V.C. Wynne-Edwards, 1962. *Animal Dispersion in Relation to Social Behavior.* Hafner Publishing Company, New York.
19. C.M. Breder, Jr. and C.W. Coates, 1932. 'A preliminary study of population stability and sex ratio of *Lebistes Copeia'.* 1932: pp.147-55.
20. W.C. Allee, A.E. Emerson, O. Park, T. Park and K.P. Schmidt, 1949. *Principles of Animal Ecology.* W.B. Saunders Co., Philadelphia and London.
21. Snyder, R.L. and Moore, S.C. 1968. 'Longevity of captive mammals in Philadelphia Zoo'. *International Zoo Yearbook,* 8: pp.175-83. Published by the Zool. Soc. Lond.
22. Lord, R.D. 1961. 'Mortality rates of cottontail rabbits'. *J. Wildl. Mgt.* 25: pp.33-40.

4 DISPERSION AND DISPERSAL

Dispersion and dispersal are treated as synonyms in the dictionary, but some ecologists use the words to mean different things. Dispersion is commonly used to define the internal spacing patterns or movements of individuals within the population. Dispersal refers to movements directed outwardly and terminating outside the geographical limits of the population. Emigrations result in population decline and immigrations produce population growth. Migrating individuals depart periodically but return later. Actually, I would prefer to think of migrations in terms of the whole population. Migratory fishes, birds, and mammals normally move collectively. It would be more precise to say that the whole population simply moves *en masse* to a new geographical dimension.

To understand why animals move, we might first contemplate why they might want to remain in one place. The problem is permanence versus mobility. Whether an animal moves any great distance or remains attached to one rather limited area is a matter of strategy. One must consider what each species requires for survival, while keeping in mind our concept of the niche. An animal's basic requirements for survival are food, water, shelter from the elements, protection from its enemies, and, in some instances, particular objects such as grit and sand for the gizzard. The last named requirement is a consequence of a particular species' unique adaptation. The species has evolved to fill a niche. The pattern of movement of individual animals and spacing patterns in the population are linked to the kind of behavior required to fill that niche.

Home Range

Most animals tend to remain within a fairly restricted home area for long periods of time. The home range of an animal is usually defined as that area traveled in its normal daily activities. Most of an animal's activity is devoted to obtaining food. Except during the breeding season, as much as 90 percent of the waking hours may be spent in gathering nutrients. Thus, larger animals have a larger home range than smaller ones, for their alimental needs are greater. Carnivores range more widely than herbivores.

The size of the home range also varies with the seasons. Wolves and foxes, for example, must range more widely during the winter when many prey animals are hibernating or protected by a blanket of snow. Conversely, deer and moose have a tendency to congregate in 'yards' no more than 100 acres in extent when weather is especially severe and threatening.

The home range of a mountain lion (puma) is fairly definite, being something of a circle, which is covered at regular intervals of three to ten days. A male lion has been known to travel over 100 square miles of habitat.

Wolves customarily assemble in packs of 6 to 30 during the winter and follow regular trails. Reported home-range sizes of individual wolf packs vary between 36 and 540 square miles.[1] There is apparently no correlation between the size of the pack and the area traveled. Packs of 16 to 20 animals on Isle Royale in Lake Superior roamed over 210 square miles. Small packs of eight to ten had home ranges between 500 and 540 square miles in Alaska and in Alberta, Canada.[2,3] There would most likely be a correlation between home range and prey density and perhaps also a correlation between range and type of prey. Wolves would have to range over a wider hunting area when prey was scarce. Their movements would also be influenced if the prey were migratory, which is the case with the caribou *(Rangifer caribou)*.

The home range of the white-tailed deer is surprisingly small, under normal circumstances no greater than one-half square mile. Most wildlife biologists in the northern regions of the United States have witnessed 'yarding' behavior of the white-tails during severe winters. The deer congregate in dense, warm coniferous stands which very often lack good quality browse. An abundance of adequate food may exist less than a mile distant, yet large numbers of these animals starve to death every winter.

It should be understood that the range of an individual animal almost always overlaps that of its neighbor. Animals are solitary, gregarious, colonial, etc., thus the home range concept applies to the individual. The eastern woodchuck leads a solitary existence while its cousins, the marmot and the prairie dog, are strictly colonial. Two deer may roam together over their home range and occupy it as a pair, yet home range is reckoned in terms of each animal separately.

In general, a typical red fox home range is one to three square miles (259 to 777 ha), the space being about one and one-half miles (2.4 km.) across at its widest point.[4] The space is usually shared by one adult male, one (rarely two) adult female(s), and seasonally by their young. Other red foxes are excluded from this space, but its borders are not patrolled. Although a mated pair shares the same space, the individuals are seldom together. They may use the same portions of the space but at different times, even though they are usually active simultaneously. Most of the fox's activity occurs at night and the remainder during late afternoon and early morning. While a fox is active, more than 80 per cent of its time is spent in travel; an individual covers a major part of the home range during the time it moves at night. Red foxes average about eight miles (12.8 km.) of travel per 24 hours and may travel as much as 15 miles (24.1 km.), all within the confines of the home range.

Some mammals, notably rodents, insectivores, and lagomorphs, establish a home range and remain there for a lifetime. Other animals will relocate their home sites from time to time, but the distances moved are not so great. Males may have a home range two or three times the size of that of the female, and different individuals of the same sex may have ranges quite unequal in size.

Calculation of Home Range Size

Areas of home ranges are calculated in various ways. Kenneth Gardner used a triangulation method (A) to determine the minimum home range of cottontail rabbits on the Conemaugh River study area in Pennsylvania. Capture sites of animals caught at three or more locations were plotted on a map and the resulting geometric figure measured with a planimeter. The size of the geometric figures varied greatly from season to season and by sex. The home range space calculated by this method varied from 0.08 to 37.2 acres for males and 0.06 to 73.6 acres for females. The mean home range space was 3.63 acres for 147 animals. Figure 14 shows the home ranges graphed by cumulative percentages.

Gardner used a second method (B) to calculate home range space for rabbits caught only two times at different trapsites. The distance between the two trapsites was considered the diameter of the home range. Home ranges calculated by this method varied from 0.3 to 419 acres for males and 0.3 to 85.3 acres for females. The mean home range was 11.8 acres for 121 animals. The graph of cumulative percentages for method B is Figure 15.

The cumulative percentage curves are plots of the frequency distribution of home range areas. These curves are skewed to the right (asymmetrical), especially in the case of method B. An inspection of the data revealed a natural break in both frequency distributions at the 14-acre class interval. The extremely large home ranges were probably measurements of seasonal movements and dispersal movements rather than true local movements on established home ranges. Thus, Table 21

Figure 14. Home Ranges of Rabbits Captured at Three or More Trapsites (Method A) (K. Gardner, Pennsylvania Game Commission)

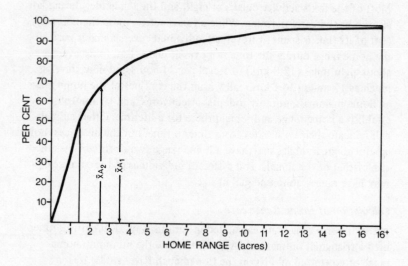

Figure 15. Home Range of Rabbits Captured at Two Trapsites (K. Gardner, Pennsylvania Game Commission)

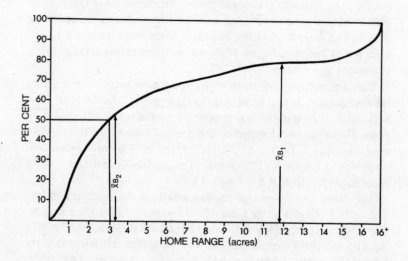

was prepared to show mean home ranges when areas above 14 acres were omitted. Mean home range was 2.53 acres using method A and 3.38 acres using method B. These means, $\bar{x} A_2$ and $\bar{x} B_2$, more nearly approached the medians or 50 percent points and were believed to measure established home range spaces.

Table 21. Mean Acreages of Cottontail Rabbit Home Ranges (K. Gardner, Pennsylvania Game Commission)

Method of Calculation	Sex	Number	Mean Home Range
A:	Males	55	2.63
	Females	64	2.42
	Weighted mean		2.53
B:	Males	61	3.90
	Females	50	2.48
	Weighted mean		3.19

The mean home ranges of male cottontails exceeded those of females. Of the rabbits caught three or more times, 53 percent had home ranges 1.5 acres or less, while only 26 percent of those caught two times had home ranges 1.5 acres or less. Over 85 percent of the rabbits caught three times or more and about 62 percent of those caught twice had home ranges five acres or less in size.

In general, this analysis of trapping data provides certain insights concerning calculation of home range space. The ideal estimate of home range would be provided by trapping the same animal repeatedly during a short period of time. Radio telemetering devices are especially effective in this respect because one animal can be followed closely for several days. However, a word of caution should be injected. A few cottontails were caught as many as fourteen times in a short period of time. Would these data furnish a better estimate of the typical home range of a cottontail rabbit? Probably not, because animals caught so frequently are not representative of normal cottontail behavior.

Territoriality

Animals defend food stores, nests, and certain areas around their home sites against trespass and exploitation by others, especially of their own kind. This is known as territoriality, and the area defended is called the territory.[5] Whatever the motivating element — food, young in the nest, drumming log, etc. — if an animal defends any part of its home range, it is said to display territorial behavior. This defensive behavior may be limited to the breeding season or to one or the other sex and usually is restricted to adult animals. This behavior pattern exists in many parts of the animal kingdom. It is especially widespread and strongly developed ritualistically in avian species and certain of the artiodactyls, particularly the African antelopes.

Dispersion and Dispersal

A distinction must be made between defense of one's person and defense of a piece of ground or real estate. There is apparently an approach distance or a zone of awareness in certain animals. *Marmota monax*, the woodchuck inhabiting the north-eastern part of the United States, leads a solitary existence except during the breeding season. Sexually mature males above ground in south-central Pennsylvania during February and March are extremely intolerant of one another. During March the distance two animals can approach before reacting to one another averages 50 feet.[6] The interaction consists of a chase, fighting, or a bluff response that involves teeth chattering and a defensive posture. During April when the newborn young are in underground burrows, the interaction distance increases to 90 feet. In May the young appear above ground for the first time, and the protective mothers are extremely aggressive and fight among themselves. The zone of awareness measures 90-95 feet at this time.

The young of the year are abandoned by their mothers toward the end of June, and they are forced to fend for themselves. At this time extensive movements commence among these youngsters which continue until at least April of the following year. The young woodchucks must remain in one place during the winter, but during the spring following hibernation a few of these animals, now yearlings, are still moving about.[7]

Young woodchucks are evidently moving about in search of appropriate home sites. The interaction distance decreases in June to less than 10 feet.[6] The testes of the males have been regressing in size since the end of March and will continue to regress until September. The seminal vesicular weight is reduced to its lowest point by June 1, indicating that secretion of testosterone has ceased by then. The lack of testosterone and the cessation of maternal agressive behavior by June 1 is no doubt responsible for the relative tolerance of the species at this time. Movements of young woodchucks continue throughout the summer months; and more and more young become involved in the dispersive movements.[7] The interaction distance increases again to 25 feet in July and 50 feet in late August and early September.[6] The interactions during the summer months are apparently initiated by the shifting young of the year. When these animals approach older woodchucks with established home burrows, they are invariably repelled. On one occasion I observed a large woodchuck kill a smaller half-grown animal that attempted to enter its burrow. Such extreme aggressive behavior is probably a rare occurrence.

Extensive trapping studies conducted on the 10,000-acre Letterkenny Army Ordnance Depot in south-central Pennsylvania indicate that most woodchucks remain for a period of time in one rather limited area.[7] Table 22 lists the mean distance traveled between successive captures for young and old woodchucks. Young woodchucks were young of the year; old woodchucks were 11 months old or older. Animals were marked with numbered ear tags for individual identification. Fifty per cent of all the movements were confined to 600 feet. A portion of the young of the year traveled distances from 3,000 to 10,000 feet between successive points of capture. Thus, the percentage of movements that were over 3,000 feet is listed in Table 23 for each group of woodchucks. It is clear that the older animals tended to remain in a small area, while some of the young woodchucks after weaning moved considerable distances.

Table 22. Woodchuck Movements — Mean Distance in Feet Between Successive Captures
(Number of movements given in parentheses.)[7]

	Area C	Area D	Area G
Old Males	490 (43)	465 (108)	390 (105)
Young Males	811 (62)	1,845 (29)	928 (65)
Old Females	373 (44)	372 (149)	421 (81)
Young Females	628 (43)	623 (33)	1,710 (26)

In order to present a clear picture of the role of emigration, movements over 3,000 feet are omitted in Table 24. Removing the long movements decreases the magnitude of the differences, but the movements of the young woodchucks were still significantly longer than those of the old woodchucks. Since the young woodchuck does not remain with its mother after weaning and since litter mates also separate during the summer, the comparatively longer movements of the young of the year

Table 23. Woodchuck Movements — Percent of Movements Over
3,000 Feet
(Number of movements given in parentheses.)[7]

	Area C	Area D	Area G
Old Males	2 (44)	1 (109)	0 (105)
Young Males	5 (62)	24 (29)	8 (65)
Old Females	0 (44)	0 (149)	0 (81)
Young Females	7 (43)	6 (33)	19 (26)

Table 24. Woodchuck Movements — Mean Distance in Feet between
Successive Captures of Young Woodchucks that did not Emigrate.
Movements Over 3,000 Feet Omitted.
(Number of movements given in parentheses.)[7]

	Area C	Area D	Area G
Males	551 (59)	527 (22)	433 (60)
Females	470 (40)	390 (31)	548 (21)

must have been due to the dispersal of the young after weaning.

With two exceptions, none of the older woodchucks of either sex
moved more than 2,600 feet, and 95 percent of their movements were
confined to 2,000 feet. A few of the adult animals were captured again
two years later at the same trapsite.

At this point we should distinguish again between dispersion and
dispersal. Dispersion covers the movements of young woodchucks away
from the maternal den after weaning. On the other hand, an emigration
is a movement out of a population. According to the definition of a
population proposed in Chapter 1, the geographical limits are designed
by the investigator. In this case, the total study area was 10,000 acres,
hence movements within the area would be considered chiefly
dispersion. Apparently the young woodchucks finally settled in one
locality, as there was no indication from these studies that older
animals roamed at random over the study area. Movements up to
10,000 feet would indicate that a small portion of the woodchucks
might be considered emigrants.

The emigration rate is generally higher in young mammals than in
adults. Young raccoons were recaptured at considerably greater
distances from the original capture site than were adults. In a study of

European rabbits, all of the 29 rabbits dispersing from warrens were young. A carefully planned study of deer mice *(Peromyscus maniculatus)* showed that young animals dispersed more than adults. Young coyotes in Yellowstone Park (US) traveled greater distances than the adults.

The European starling has spread over the entire United States and much of Canada within a period of sixty years. Attempts were made to establish these birds at West Chester, Pennsylvania before 1850, at Cincinatti, Ohio in 1872-73, and at Portland, Oregon in 1889, but all of these birds gradually disappeared. The permanent establishment of the starling dates from April 1890, when eighty birds were released in Central Park, New York City. In March of the following year, eighty more were released. In about ten years the birds became established in the New York City area.[8] This rapid extension of the population range was due to the irregular migrations and wanderings of nonbreeding juvenile birds one and two years of age. Adult birds typically use the same breeding area from year to year and thus do not colonize new areas.[9]

Migration

Migration is a more or less continuous and direct movement from one locality to another with a return to the original area. Such movements are usually co-ordinated with environmental events, especially periodic changes in temperature and food supply. Animals exhibiting migration, in the strict sense of the word include the monarch butterfly, herring, salmon, trout, eels, certain whales, fur seals, mule deer *(Odocoileus hemionus)*, African ungulates, and a great many kinds of birds.

Birds especially follow rather circumscribed paths of migration and the majority of them migrate at night. Such birds commonly travel from 200 to 400 miles per night and recuperate and feed during the daylight hours. Thus, all along the flyways the avian species are constantly impinging upon the resident population of animals. Predaceous migratory birds prey on the local non-migrating organisms, while the resident predators catch migrating avian prey. Thus, population dynamics may often be affected by migratory behavior of various animals.

Homing

The ability to return to a home area after being away for a considerable length of time or to return to the home area through unfamiliar territory is termed homing. Apparently, mammals have this ability to a greater or lesser degree. I was involved with a program of trapping and transferring white-tail deer once, and recorded marked animals returning to their home sites from distances up to 48 miles. Deer released within 12 miles of the home area returned within a few days. Similarly, several mice *(Apodemus sylvaticus)* in a study in England returned to their homes after being released on the opposite side of a lake, far beyond their normal range. Unfortunately, proving these returns are not random requires a considerable quantity of data.

The fishes may have the best developed homing 'instinct'. The spawning migrations of salmon and trout are unique in this respect, as they consistently find the place where their parents spawned after swimming thousands of kilometers from the ocean. For a long time it was argued whether these fish actually could sense their home waters or found them accidentally. In 1939 in British Columbia, 499,326 young salmon were marked and released in a tributary of the Fraser River. Years later nearly 11,000 of these fish were recovered as they made their way to the spawning sites. With minor exceptions, they ascended the same river from which they had descended in moving out into the ocean.

Fishes apparently have an internal clock mechanism and the ability to direct their swimming direction on the basis of the sun. Salmon evidently navigate much like men at sea (with clock and sextant). More is understood about migrating salmon when they reach fresh water. Laboratory studies have shown that salmon and trout have a well-developed olfactory sense and react to all kinds of scents. A field study of silver salmon in two arms of the Issquah River revealed how scents can help fishes find their home waters after years in the ocean. The nasal pits of one group of silver salmon were stuffed with cotton, while another group of fish acted as controls. Salmon breathe through their gills, so plugging the nostrils would produce no discomfort or interference with oxygen consumption. These fish were taken further downstream and released. Only the fish with unstopped nasal pits found their way back to the home water. The salmon with stopped-up nasal pits descended the two river branches randomly. We must assume that salmon and trout have a good memory of the faint characteristic odors of the spawning beds where they were hatched.

Dispersal and Population Density

The success of immigrants in establishing home sites depends chiefly upon the local population level. The immigration of rats into city blocks is negligible when the populations have reached carrying capacity and trivial even in growing populations.[10] The widespread failure of stocking programs for game species is probably related to the intolerance of the resident conspecifics that are already existing at carrying capacity. Stocking programs may be successful if the new habitat is devoid of competitive species. Strecker[11] found that emigration was negligible in a colony of house mice until the limited food supply was being utilized to capacity. Then the number leaving the area increased greatly (Figure 16).

Figure 16. Increase of the Number of Emigrants From a Mouse Population With Population Growth[11]

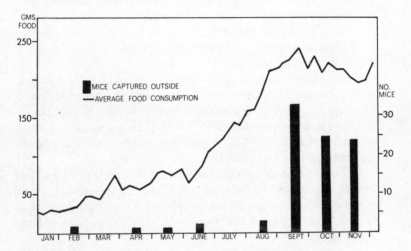

In the discussion of mortality, mention was made of increased vulnerability to predators and other causes of death of animals in strange places. Dispersing animals are in unfamiliar surroundings. Indeed, a major explanation for the tendency of animals to remain within a home range is the increased protection afforded by familiarity with surroundings.[12] Paul Errington[13] noticed the severity of losses of muskrats traveling from one pond to another. He was of the opinion that numbers of muskrats were determined by the hostility or intolerance of muskrats toward one another, and that muskrats driven out of their familiar home ranges or from natal ranges in some instances were doomed to die — if not from predators, then from disease and exposure.[14] Thus, social interaction and population pressures are invoked repeatedly to explain why various biological events are related to population density.

Summary and Conclusions

The force for dispersal may result in an increase or a decrease in the population depending upon its net value. The natural rate of increase can be written as the sum of births and immigrations less the sum of deaths and emigrations. The pattern of movement of individual animals and the spacing patterns within the population depend upon adaptive behavioral characteristics and elemental requirements for food, water, and shelter.

Most animals tend to remain within restricted areas of the habitat for long periods of time. Familiarity with surroundings probably has survival value, which might explain why most animals remain in one area and also why territories are established and protected in many instances. Territoriality and intraspecific intolerance within the home range space are effective mechanisms for population dispersion and regulate or limit population density as well.

Young animals are evidently forced to move considerable distances to find unoccupied ranges or unprotected territories. Inexperienced migrating animals in unfamiliar environs are vulnerable to predation, accidents, and unfavorable weather conditions which increase their probability of dying (q). Colonization of unoccupied habitats is largely achieved by the younger members of the population.

Characteristic mean population densities are determined by several factors: food habits (e.g. predator − solitary or co-operative), position in the food chain, species behavioral patterns (e.g. solitary, colonial, herding, flocking, social), age at sexual maturity, reproductive potential, and longevity.

Migratory behavior is a special adaptive characteristic of many species of insects, birds, and mammals that enables them to exploit habitats that would be otherwise inaccessible. Salmon and trout migrate into the ocean to complete their growth where food supply and space are nearly unlimited. The smaller inland streams and rivers would be unable to support the potential biomass of these species. Migratory birds by spending their breeding, nesting, and brooding seasons in the great subarctic and arctic spaces are able to exploit vast regions of the earth. Since the same regions are inhospitable in the wintertime, migration to more favorable climates is required for survival.

The success of immigrants in establishing home sites depends chiefly upon the local population level. Carrying capacity is a useful concept inasmuch as population density is dependent upon the existing requisites of the habitat and the inherent behavioral characteristics of the species with respect to its utilization. Dispersal force is an important aspect of population dynamics. The overall integration of this force with the forces of reproduction and mortality will be discussed in the following chapter.

Notes

1. L.D. Mech, 1966. *The wolves of Isle Royale,* Fauna of the National Parks of the United States, Fauna Series 7, Supt. of Documents, US Government Printing Office, Washington, DC 20402.
2. B.L. Burkholder, 1959. 'Movements and behavior of a wolf pack in Alaska'. *J. Wildl. Mgmt.* 23: pp.1-11.
3. W. Rowan, 1950. 'Winter habits and numbers of timber wolves'. *J. Mammal.* 31: pp.167-9.
4. G.G. Montgomery, 1974. 'Communication in red fox dyads: a computer stimulation study'. *Smithsonian Contributions to Zoology,* No. 187, 30 pp.
5. W.H. Burt, 1943. 'Territoriality and home range as applied to mammals'. *J. Mamm.* 24: pp.246-352.
6. F.H. Bronson, 1961. Doctoral Dissertation, Dept. Zoology, Pennsylvania State University.
7. R.L. Snyder, 1960. 'Physiologic and Behavioral Responses to an Altered Sex Ratio of Adults in a Population of Woodchucks'. Johns Hopkins Univ., School of Hygiene and Public Health.
8. M.T. Cook, 1928. 'The spread of the European starling in North America'. US Dept. Agr. Circ. No. 40, 9 pp.
9. B. Kessel, 1953. 'Distribution and migration of the European starling in North America'. *Condor* 55: pp.49-67.
10. J.B. Calhoun, 1948. 'Mortality and movement of brown rats in artificially supersaturated populations'. *J. Wildl. Mgt.* 12: pp.167-72.
11. R.L. Strecker, 1954. 'Regulatory mechanisms in house mouse populations: The effect of limited food supply on unconfined population'. *Ecology* 35: pp.249-53.
12. D.E. Davis, and F.B. Golley, 1963. *Principles in Mammology.* Reinhold Publishing Corporation, New York, Chapman and Hall Ltd., London.
13. P.L. Errington, 1946. 'Predation and vertebrate populations'. *Quart. Rev. Biol.* 21: pp.144-77.
14.———1963. *Muskrat Populations.* Iowa State University Press, Ames.

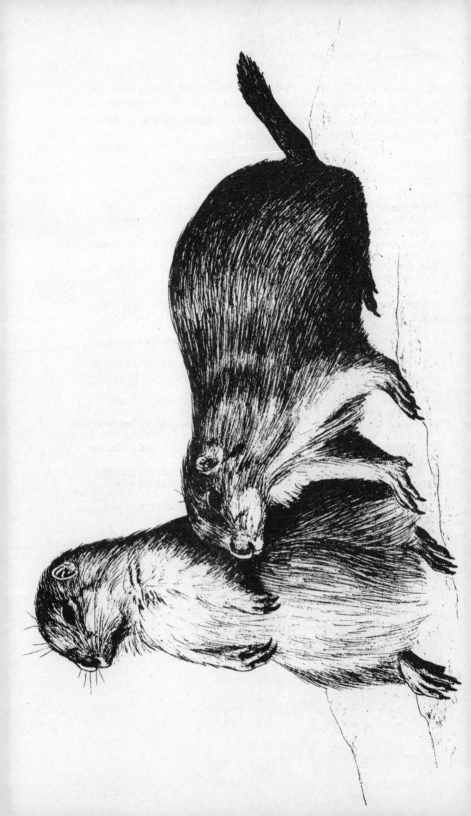

5 POPULATION DYNAMICS

The study of populations is inevitably a quantitative one, since the most important ecological problems involve numbers. Mankind is constantly concerned about the sizes of populations of plants and animals: wheat yields, cattle populations, game bird populations, vermin populations, predator populations, mosquito populations, etc. The problems are basically why a population is as large as it is and why it changes in size. Such questions involve kinetics and equilibrium, hence the study of population dynamics. Generally, the analysis of biological populations is undertaken for two reasons: (a) beneficial species are managed to increase population size for exploitation and (b) detrimental species are managed to decrease population size.

Studies of population dynamics require enumeration and observation of changes in time. The changes in population density with time are correlated with data collected on reproduction, mortality, and dispersal to answer the two fundamental questions: why a population is as large as it is and why it changes in size. For the first time, we are considering all population forces simultaneously. The practical application of such knowledge allows management of populations.

Law of the Minimum

The first experiments conducted by students in the college laboratory to demonstrate principles of population dynamics utilize bacteria and test tubes. If six different test tubes of equal size are provided with equal amounts of ingredients (a nitrogen source, certain salts, a phosphate buffer and glucose) and equal numbers of bacteria *(Escherichia coli)* are introduced, the populations of bacteria will all stop growing at exactly the same time when they are incubated at 37°C.

If one of the ingredients is supplied in graded quantities, the results are quite different. Figure 17 illustrates the population growth curve if glucose is provided in graded concentrations as follows: Test tube no. 1 − 4.0 gram/litter, no. 2 − 2.0 g./1., no. 3 − 1.0 g./1., no. 4 − 0.5 g./1., no. 5 − 0.25 g./1., and no. 6 − 0.125 g./1. In this case the population density of *E. coli* is measured by turbidity. The higher concentrations of bacteria are indicated by increased turbidity. In the schematic population curve represented by Figure 17, the turbidity was measured in a densitometer, an instrument that measures the amount of light coming through the liquid in the test tube. The test tubes are gently vibrated to insure uniform dispersion of the bacteria in the medium. As the number of bacteria increase, less light is transmitted to a

Figure 17. Growing Populations of Bacteria Related to Glucose Concentration

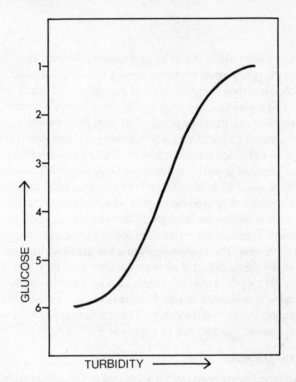

photoelectric cell in the densitometer. The bacteria could have been measured more directly with certain bacteriological counting techniques, but the procedures would have been complicated and time-consuming. The problem of direct enumeration is a complicated, time-consuming process in nearly all population studies, hence measures of relative densities as in the present experiment are used in many cases.

The growth form of bacteria under these circumstances illustrates one of the first laws of population growth, often called the law of the minimum. A German scientist, Justus Liebig, is credited with the discovery of this law in 1876. Stated simply, the law is that population growth will be restricted by the factor in least amount. Just to insure that glucose was the limiting factor in each test tube population, a small amount of glucose, 0.004 grams was added to each tube. When the test tubes were incubated again at 37°C, growth of bacteria commenced again in every tube except tube no. 1. Duplicate tubes or controls for each experimental test tube were also incubated again

without the addition of glucose to insure that re-incubation was not
the stimulus for renewed population growth. This simple precaution
illustrates the scientific method of inquiry — that nothing should be
taken for granted.

Chemical Regulatory Factors

The failure of the population in test tube no. 1 to grow again after
more glucose was added points out another principle of population
growth. The experimental populations growing in test tubes of limited
capacity will reach a certain size and no more. Further population
growth is possible only when the media is renewed. Thus, bacterial
populations may be limited ultimately by toxic products of metabolism.
Analogous inhibiting substances, not necessarily excretory products,
are produced in nature by at least two species of snails, *Fossaria
cubensis* and *Biomphalaria glabrata*, flour beetles, and frog tadpoles.[1,2]
A chemical substance produced by *F. cubensis* can inhibit its own
species growth and egg deposition.[1] An ammonium sulphate precipitate
prepared from old cultures of *B. glabrata* has an inhibitory effect on
young *B. glabrata* and on egg production by the adults.[1] A proteinaceous
growth inhibitor is released by frog tadpoles.[2]

Certain species of flour beetles have paired odoriferous glands in the
thorax and abdomen that produce an irritant gas when stimulated by
mechanical disturbance, crowding, and chilling.[3,4] The substance, mainly
ethylquinone, is lethal in crystalline form when applied to first-star
larvae.[5,3] In gaseous form it induces developmental abnormalities in
late larvae and pupae.[5] Many population principles were derived from
extensive studies of flour beetles earlier in the century. Contamination
of the flour, which serves simultaneously as home, shelter, and food
for the beetles, eventually prohibited further population growth. In this
case, ethylquinone may have been the antaphrodisiac or
fertility-depressing substance that accumulated in the flour to reduce
reproductive rate at late stages of population growth.

Allelopathy

Antibiotic interactions between plants are caused by products of
metabolism. This influence of plants upon one another is called
allelopathy. Interest in toxic secretions of plants was stimulated by
observations recorded in the nineteenth century. When one crop was
grown continuously on one plot of ground, the yield decreased and
could not be increased by application of fertilizers. Many fruit trees
will grow poorly if planted in an orchard where the same kind of fruit

tree had grown previously. Subsequently, a great many examples of
'soil sickness' have been discovered. For example, grass has a
detrimental effect on apple trees;[6] walnut trees secrete a toxin
(probably 5-hydroxy-a-napthaquinone) to which some plants are
susceptible;[7,8] barley inhibits the germination and growth of several
weeds;[9] peach trees succeeding apples grow better than peach trees
succeeding peaches;[10] and the parent trees *(Grevillea robusta)* in the
subtropical rain forest of northern Australia kill their own seedlings
chemically.[11]

Unlimited Population Growth

Although the growth of any population must eventually be limited,
a study of simple unlimited population growth is a helpful preliminary
to understanding limited growth. Bacteria reproduce by binary
fission, thus a population curve for *E. coli* can be graphed from the
following data:

Time	0	1	2	3	4	5	6	7	8
Number	1	2	4	8	16	32	64	128	256

From these data we can see that the number at time t is 2^t. The growth
rate (r) is the natural logarithm of 2 or $\ln 2$. The number at time t
is $(e^r)^t = e^{rt}$, where e is the base of natural logarithms, approximately
2.718. If we start with any other number of bacteria (N_0), then all the
numbers above will be multiplied by N_0. Thus, the formula for unlimited
population growth can be expressed:

$$N_t = N_0 e^{rt},$$

This formula can be used in various ways. For example, consider a
population that increases from 150 animals in 1960 to 170 animals in
1970. Taking the natural logarithm of each side of the equation (and
remembering that $\ln e = 1$) gives:

$$\ln N_t = \ln N_0 + rt$$

$$r = \frac{\ln N_t - \ln N_0}{t}$$

$$r = \frac{\ln N_t/N_0}{t}$$

$$r = \frac{\ln \frac{170}{150}}{10} = \frac{\ln 1.13}{10} = 0.01222.$$

Also, given the growth rate, the doubling time of a population can be computed by setting $N_t/N_0 = 2$. Then $2 = e^{rt}$. Taking the natural logarithm of each side gives:

$$\ln 2 = rt \text{ or } \frac{\ln 2}{r} = t \text{ or } \frac{0.6931}{r} = t.$$

For the population above with a growth rate of 0.01222, the doubling time is 56.7 years.

If we plot n^t against t, we will get a curve of increasing slope as in Figure 17. If we plot $\ln n$ against t, it will be a straight line with slope r. Growth of this sort is unlimited, exponential and autocatalytic — unlimited because n always increases with increasing t, exponential because of the term e^{rt}, and autocatalytic because the rate of growth depends on the number of animals already present in the population.

Limited Population Growth

Figure 18 records a confined population of nutrias *(Myocastor coypu)* over a period of twelve years in the Philadelphia Zoo. This is a typical example of population growth when space is limited and dispersal is prevented. Such populations often become extinct as this one did in 1958. The nutria pen was roughly circular, about 10 meters in diameter, and contained a pool and stone shelter. Productivity was indicated by offspring that survived for one month after birth. Any that survived from one calendar year to the next were listed as adults.

The first part of the growth curve is nearly a straight line and corresponds to the formula for exponential population growth. The rate of increase calculated by the formula $N_t = N_0 e^{rt}$ is 0.71555.

The nutria population was started with an adult male and female. More than half of all offspring died after 1949 within a month of birth. Only ten offspring were produced after 1951 and four of these survived to adult age. None of the four reproduced. The population crash occurred because of increased mortality and inhibition of reproduction. Since food and water were supplied in excess of needs, decline in numbers was attributed to endocrine responses to increased social interactions.

A group of prairie dogs *(Cynomys ludovicianus)* was established in a small outdoor enclosure in the Philadelphia Zoo in June 1954 with 24 individuals. Cropping of 5–10 prairie dogs every year prevented the population collapse, and reproduction continued year after year. Mortality was still a factor but appeared to be independent of density. Subsequently in 1965 cropping was discontinued to study a complete population cycle. The population declined from approximately 48 animals to less than 11, when new animals were added to the enclosure to rejuvenate the population. Apparently social interaction at high population densities triggers the pituitary-adrenal-gonad responses that inhibit reproduction. Decreased resistance to infectious disease, a stress response, may also contribute to mortality. Removal of surplus animals periodically seems to alleviate social conflicts and prevents extinction.

Figure 18. Population Curve of a Confined Population of Nutria *(Myocastor coypu)*[12]

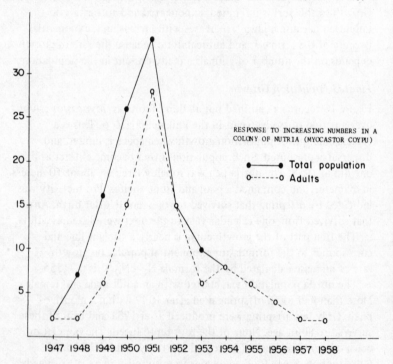

RESPONSE TO INCREASING NUMBERS IN A COLONY OF NUTRIA (MYOCASTOR COYPU)

●——● Total population
○------○ Adults

The Logistic Curve

When there is density-dependent limitation, there will be a relationship between the rate of change of the population and the population size. Many populations when introduced into a new locality show what is known as sigmoid population growth. When population numbers are plotted arithmetically against time, the curve is s-shaped; the rate of increase is slow to start with (dn/dt is small), it increases rapidly (the logarithmic phase) at greater population size (dn/dt larger) and flattens off to an asymptote eventually (dn/dt = 0). The most popular mathematical expression for fitting data to this curve is known as the logistic equation.

The logistic equation is a simple elaboration of one form of exponential growth.[13] Differentiating $N_t = N_0 \, e^{rt}$ gives d ln n/dt = r. The logistic equation is the simplest density dependent elaboration of this:[13]

$$\frac{d \ln n}{dt} = r - c\,n,$$

where r is, as before, the intrinsic rate of natural increase, and c is a constant. The equation is not usually written in this form, even though it is the simplest.[13] The usual form is derived as follows:

$$dn/dt = n(r-cn) = rn - cn^2 =$$
$$rn\,(1-cn) = rn \; \frac{k-n}{k} \, .$$

This introduces a parameter K which is 1/c. In this differential form there are two parameters, r which gives the slope of the initial part of the curve, and K which will be found to be equal to the equilibrium value of n. It is important to note that K is an upper asymptote, a balance point and not a limit to the population.[13] That is to say if n is artificially increased above K, the population will decline to K. The third parameter comes in when the logistic is written in integral form.

$$n_t = \frac{K}{1 + \exp{(z-rt)}}$$

where z is a constant involving n_0. Again, in the usual convention, both r and c (or K) include both births and deaths. The integrated form of the logistic equation is also written as follows:

$$N = \frac{K}{1 + e^{a} - rt}$$

where:

   N = population size
   r = specific growth rate
   K = maximum number possible or upper asymptote (carrying capacity)

e = the base of natural logarithms

a = r/K.

The logistic equation can be written in yet another way by taking the natural logarithm of each side and rearranging the terms:

$$\ln N = \ln \frac{K}{1 + e^{a - rt}}$$

$$\ln K = \ln N + \ln N(a-rt)$$

$$\ln \frac{K-N}{N} = a - rt.$$

This is the equation of a straight line in which the co-ordinates are

$$y = \ln \frac{K-N}{N}$$

$$x = t$$

and the slope of the line is r. This relationship can be used to fit a logistic equation to actual biological data.[14]

The logistic curve was first suggested to describe the growth of human populations in 1838.[15] The same equation was independently derived by Pearl and Reed[16] in 1920 as a description of the growth of the population of the United States (see Figure 19).

Equilibrium Densities

Thus far, only the growth phase of population dynamics has been analyzed, but unless the species is truly cyclic, pioneer populations will eventually reach an equilibrium size within the community. That is not to say that population levels won't change again. Environmental factors are constantly changing, and community populations are continuously being adjusted to one another and to the changing environment.

The rabbit population on the Conemaugh River study area graphed in Figure 20 is representative of established populations. The population each year was measured in the fall. SE in Figure 20 is the standard error of the population estimate. Statistically this figure means that the true population level will fall within ± two standard errors 95 percent of the time. Five percent of the estimates will fall outside these limits. The population estimates may not seem very precise, but this is true of most techniques for estimating population size. Actually, the change in population from 1957 to 1959 was statistically significant,

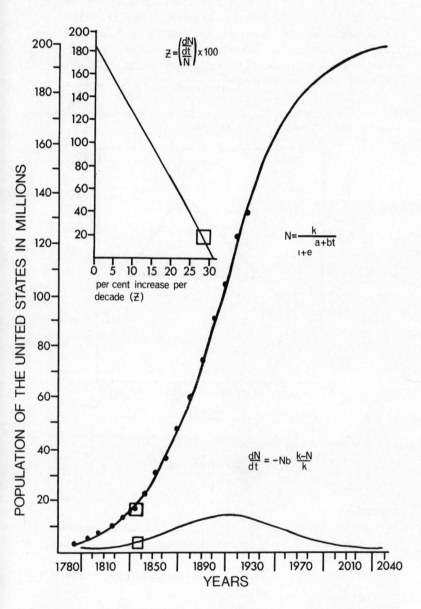

Figure 19. The Logistic Curve of Increase of Population in the United States (The insert shows the inverse relation of population gain to population size.)

which means that mathematically a claim can be made for a real change in population levels. In this instance the study area was being managed to increase rabbit numbers by planting grasses, clovers, and other forage plants, producing brush piles, and planting shrubs and pine trees for additional cover. Thus, it would be desirable to be able to claim a valid population response to the management practices. An adjacent unmanaged study area was used as a reference or control area.

Figure 20. Calculated Rabbit Populations ± 1 SE on the Conemaugh River Study Area (K. Gardner, Penna. Game Commission)

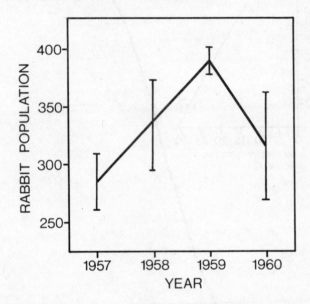

The technique used to estimate population levels is called the recapture method or Petersen index. This formula is:

$$N = \frac{Mn}{m}$$

where:

M = rabbits marked and released during the first trapping period
n = total captured during the second trapping period (marked and unmarked)
m = marked rabbits captured during the second trapping period

$$SE = \frac{M^2 n (n-m)}{m^3} .$$

Social Organization of Confined Mouse Populations

With so much emphasis on numbers, we sometimes lose sight of the
activities and the fates of individual members of the population.
Figure 21 is a graph showing the basic population dynamics of house mice
confined in the Penrose Research Laboratory. The population
represented was the first of several studied in this laboratory between
1960 and 1966. Food and water were supplied daily in excess of needs,
and shelter cans and nesting material were added continuously as
these populations increased in size.

Figure 21. Population Dynamics of Wild House Mice Confined to 43ft.[2] of
Floor Space [17]

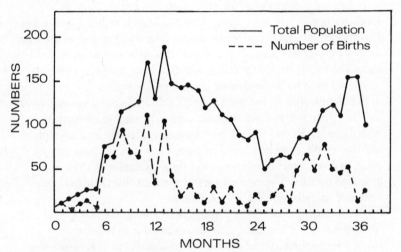

One can see at a glance that reproductive rate was inversely
proportional to density in this population. At first, growth was slow
because only two females were introduced to start the population, but
as the young of the first two litters matured and reproduced, the
number of young born each month increased precipitously. However,
the population ceased to grow after reaching a peak of almost 200 in
13 months. Numbers decreased slowly over a period of 12 months
until 50 mice were left. A new surge of reproduction started the
population on a new growth phase that lasted 11 months at which
time the second gradual decline commenced.

Most experimental mouse populations have been terminated soon
after the asymptote was reached. In this respect the population
described here was unique as it was finally terminated in 1965 after

reaching a third peak density. Three other populations of house mice studied in similar cages reached asymptote densities in 11 or 13 months, and each showed the same gradual decline over the next 12 months.

Crew and Mirskaia[18] were the first to recognize that population pressures affected certain animals more than others. Clark[19] described this phenomenon in considerable detail in his studies of the vole, *Microtus agrestis*. He distinguished three classes among the males: (a) large, glossy-furred, relatively unscarred individuals that roamed with impunity throughout the enclosure; (b) similarly large, free-roaming individuals, but with tattered fur and scars on the hindquarters; and (c) lightweight unaggressive, tattered, lackluster individuals with restricted home ranges.

I observed a similar state of affairs among the male house mice confined to 42ft.[2] of floor space in double-decker, stainless steel cages. The mice could gain access to the second floor by climbing up the wire mesh walls on three sides. These populations were started with two males and two females of breeding age. One male was invariably killed by the other within two or three days. During the first year of population growth, the mice moved freely throughout the enclosure and the females established nests for their offsprings in all parts of both decks. However, as these populations reached the first asymptote and the reproductive rate began to drop, a remarkable change took place. Several classes of mice, in some respects similar to those described by Clark,[19] now were clearly apparent. Furthermore, these classes of animals were spatially separated.

The following classes were distinguished among the males: (a) dominants — a few glossy furred, unscarred individuals that roamed with impunity throughout the enclosure, but that rested and fed on the first floor; (b) recluses (Figure 22) — the cage was not designed with this in mind, but the upright supports, stopping short of the roof, allowed a small space on which a mouse could perch. The position seemed uncomfortable, but certain males called recluses occupied these sites during the daylight hours. Populations were not observed at night, but because longevity was not affected significantly, these animals apparently came down periodically to obtain food and water. (c) Huddlers (Figure 23) — a third class of males huddled together in aggregations usually located in a corner of the second deck. These groups on occasion contained more than 100 individuals of both sexes. (d) Withdrawn individuals practically hairless, with many suppurating wounds — these males did not huddle and were often observed on top of a shelter can, but rarely inside one.

Figure 22. Recluse Perching on Upright Support of Cage

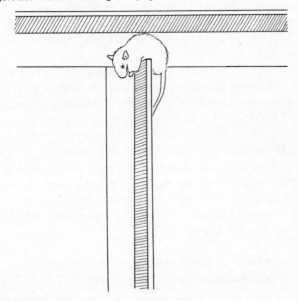

Figure 23. Mice Huddled in Corner of the Second Deck

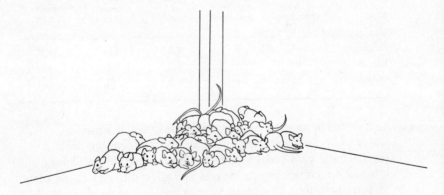

Females were characterized as follows: (1) a few, large, well-furred, unscarred individuals restricting their movements to the first floor — these females were responsible for the small number of births that occurred during the periods of maximum density and the post-asymptotic declines; (b) a rather indistinct class — lightweight, relatively unscarred individuals occupying shelter cans in the second deck either alone or with one or two other females. These females produced offspring in these shelters, but during periods of high density or during the periods of declining density these newborn pups invariably perished, usually within 24 hours of birth; and (c) comparatively lightweight, relatively unscarred individuals that joined the large huddles on the second deck.

Another measure of the effects of population pressures on different social classes was obtained by testing semen characteristics of various males at peak densities. Data in Table 25 show characteristics of semen collected from males representing three classes — dominants, recluses and huddlers. Statistical treatment of these data showed no significant differences between classes, either in success in stimulating ejaculation or in seminal volume and sperm concentration. However, the average sperm concentration and average number of spermatozoa in ejaculates collected from dominant males were clearly within normal ranges, whereas ejaculates collected from recluses and huddlers were well below the normal ranges for these values.

Table 25. Characteristics of Semen Collected by Electroejaculation From Mice in Confined Populations at Peak Density[17]

Class	No. stimulated	Semen success-fully collected	Mean vol. (mm^3)	Mean sperm conc. $(1000/mm^3)$	Mean No. sperm (1000)
Dominant	10	7	8.8	434	2498
Recluse	10	5	9.3	188	1479
Huddler	10	3	9.8	68	537

It should be apparent by now that the relationship between reproduction and density is a highly complex one involving both psychological and physiological factors. In the realm of psychology, it is now certain that population pressures affect different individuals to different degrees. This brings to mind these questions: what determines how an individual will react to population pressures? Does the individual's response depend upon genetically endowed or acquired characteristics?

Social organizations involving one or two dominant individuals and several subordinates are termed dominance hierarchies. Many investigators believe that characteristics or traits that confer dominance are genetically determined. Wynne-Edwards[20] uses the hierarchy as an example of a typical group character, manifested in a collective group but absent in individuals and therefore subject to group selection. He visualizes the dominance hierarchy as a mechanism for regulating population density. Wiens[21] discusses this problem at length in his paper on group selection and Wynne-Edwards' hypothesis. In general, his conclusions are these: 'If a variety of features can combine in various ways to confer dominance, then, through the reproductive advantage which often goes with dominance status, this variety will be passed on to subsequent generations by simple individual selection.'

Ecologists and behaviorists have speculated about what characteristics or traits would be advantageous to the dominant animal. Are strength, sharp teeth, large antlers, or intelligence some of the favored traits? One can also not help but wonder about the natural selection of the dominance status. If the dominant individual has a reproductive advantage, would it be reasonable to assume that sooner or later all individuals would have these dominant traits? Then, if this were so, which individuals would be subordinate? Is the alternative, proposed by Wynne-Edwards that the hierarchy has arisen through group selection, the only solution to this problem?

Interpretation of data from the freely growing population of brown house mice suggests that the temporal circumstances of birth may confer the dominant status. In the population of wild mice it was not so much size and sex that conferred dominance but rather the fortunate circumstance of being born when population density was lowest. This conclusion is supported by an analysis of the data on longevity. Newborn mice were marked for individual identification which made it possible to determine age at death of each mouse. Figure 24 illustrates graphically the mean age at death of mice born at various monthly intervals. The number of births each month is given in Figure 21. For the most part the average longevity of mice born during the first five months of the population's existence was much greater than that of mice born subsequently. Initially only one or two litters were delivered each month involving only five to ten offspring. With the exception of those members of a litter born during month 4, the mice born to the founding females lived between 200 and 325 days on the average. Also, approximately 60 mice born during month 5 (F_1 generation progeny now included) lived on the average slightly more than 200 days. Thereafter, average longevity was less than 100 days and more often than not newborn perished within 24 to 48 hours.

Continuing with this line of investigation, it was found that dominant

Figure 24. Mean Life Span in Days (Ordinate) Related to the Month of Birth (Abscissa)[17]

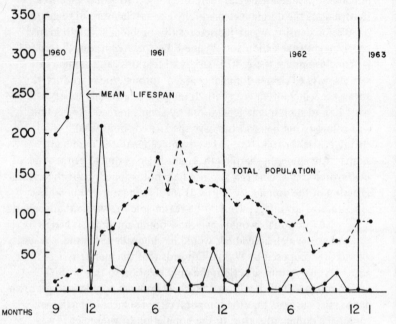

males came from those litters delivered during the first five months of population growth. Also recluses, always relatively few in number, were more often than not from the early litters. The same situation held for the dominant females.

Thus, it would appear that the role of inheritance in conferring dominance was not very important. The population originated from two females and one male. There was no indication that the offspring of one female possessed any advantage over the offspring of the second. Dominance was entirely a matter of the temporal circumstance of birth.

Terman[22] found a similar state of affairs in freely growing laboratory populations of prairie deer mice founded by releasing four adult males and four adult females into enclosures containing food and water in excess of utilization. 'Control' of population growth was achieved by either one of two means — cessation of reproduction or failure of the young to survive. Reproduction of founding females was significantly greater than that of females born in the enclosures. Slightly more than 37 percent of the founding females reproduced, but only 6 percent of the females born in the enclosure reproduced subsequently. Approximately 83 percent of females alone in small breeding cages produced young while these populations were studied. These experiments again illustrate the inverse relationship between density and reproduction, but more important perhaps was Terman's discovery of the reproductive advantage conferred to the founding females.

Wien's[21] concept of the dominance status, especially with respect to age, deserves additional comment. He says it is well known that in many species age generally confers higher social status. This statement may be true as far as it goes, but the problem involves the matter of continuity. When the population in question was studied, it may well be that older animals were dominant at that particular time, but how long had the dominant animal enjoyed this high rank? Was it achieved later in life or did the animal enjoy this advantage from birth onward? Wiens[21] further states that the features that confer dominance are not fixed; any subordinate individual, although initially excluded from breeding, may later attain a position of relative dominance. Under the circumstances, it would seem desirable to design experiments and studies to bear out such positive statements, because I am not convinced that the 'features' that confer dominance are so flexible or that it is age *per se* that confers the higher social status. Convincing evidence, either pro or con, would require observing individuals over their entire life span under a variety of circumstances.

Additional Formulas

There are only four factors that can change the size of a population:

$$B = \text{births}$$
$$D = \text{deaths}$$
$$I = \text{immigrants}$$
$$E = \text{emigrants.}$$

These factors can be combined to form a fundamental equation for population dynamics:

$$N_{i+1} = N_i + B - D + I - E.$$

This fundamental equation can be adapted to several situations if constants are added. Thus far, B, D, I, and E have been considered as factors dependent on N but any number of interactions with environmental agents can be included in the equation. For example B can be related to quantity of food if the strength of the interaction can be determined, or D can be made age-specific. The complexity of the formula is naturally increased by such additions.

In nature there are time lags between environmental changes and their corresponding effects on population growth. Reaction time lags can be incorporated into the regulation $[(K-N)/K]$ of the logistic equation as follows:

$$\frac{dN}{dt} = rN \frac{K-N_{(t-a)}}{K}$$

where:

a = reaction time lag.

A reproductive time lag, which can be measured by the gestation time or its equivalent, can be incorporated into the logistic equation as follows:

$$\frac{dN}{dt} = rN_{(t-g)} \frac{K-N_{(t-a)}}{K}$$

where:

g = reproduction time lag.

The interesting thing about time lags is their effect on population growth patterns. The simple addition of time lags into population models causes oscillations in numbers in which the population continuously overshoots and then undershoots the equilibrium density.

The Lotka-Volterra equations describe competition between species for food and space in terms of the logistic equation. For example, one equation can be written for species 1 and a second for species 2. If the species are interacting, thus affecting the population growth of each other, a conversion factor is added that expresses species 1 individuals as an equivalent number of species 2 individuals in terms of competitive interaction:

$$N_1 = \alpha N_2.$$

This is a simple conversion factor that ignores density effects. The Competition equation for species 1 is written:

$$\frac{dN_1}{dt} = r_1 N_1 \frac{(K_1 - N_1 - \alpha N_2)}{K_1}$$

also:

$$N_2 = \beta N_1$$

and:

$$\frac{dN_2}{dt} = r_2 N_2 \ \frac{(K_2 - N_2 - \beta N_1)}{K_2}$$

With interacting species only three results are possible: (a) both species coexist, (b) species 1 becomes extinct, or (c) species 2 becomes extinct.

Simple models for predator-prey interactions can also be formed with logistic equations. First a logistic formula is derived for the prey population in the absence of predation. Then various parameters describing predator efficiency must be considered. The predator may take a constant percentage of prey or all of those not protected in safe cover. If predator pressure is density-dependent the whole predator population must capture proportionately more prey when they are abundant and proportionately less when they are scarce.

Thus:

$$N_{t+1} = \ \ (1.0 - B_{xt}) N_t - E N_t P_t$$

where:

t = generation number
B = slope of reproductive curve
x_t = $(N_t - Neq)$ = deviation of present population size from equilibrium population size
P = population size of the predator
E = a constant measuring the efficiency of the predator.

If the reproductive rate of the predator depends on the abundance of the prey, a constant measuring the efficiency of the utilization of prey for reproduction by predators is added.

Then:

$$P_{t+1} = Q N_t P_t.$$

Note that if the prey population were constant, this equation would describe geometric population growth for the predator. When predators are absent and prey scarce, the reproductive rate of the prey will be approximately:

$$N_{t+1} = (1.0 - BNeq) N_t.$$

When the prey population is at equilibrium, the few predators present will increase at

$$P_{t+1} = Q \, Neq \, P_t.$$

These are maximum reproductive rates for the prey and the predator. Calculations are tedious but population sizes for both prey and predator can be determined for an unlimited number of generations and with different predator efficiencies.

There are four possible outcomes: (a) stable equilibrium with no oscillation, (b) stable oscillation, (c) convergent oscillation, and (d) divergent oscillation leading to the extinction of either predator or prey.

All equations considered thus far were so-called deterministic models in which each response was predicated on a pre-existing set of conditions. Stochastic models employ a set of probabilities for each interacting factor which make such models more realistic in terms of natural population dynamics. Stochastic models are best handled with computers and can be exceedingly complex.

Summary and Conclusions

The growth of a population can be described with simple mathematical models that sometimes correspond to the realities of population dynamics in the laboratory. Several interacting factors can be formulated: the concept of carrying capacity, ultimate limits to population growth, inhibitory factors, allelopathy, reproductive potential, density-dependency, instantaneous rate of increase, equilibrium densities, and competitive interaction.

The one item missing in both deterministic and stochastic models of population growth is a constant for the variable actions of individual members of the population. Mathematical models assume likeness of individuals which is also the definition of population presented in Chapter 1. The larger metazoans of the same species are seldom very much alike, but differ according to their own genetic constitution, their early experience and opportunities for learning, and the temporal circumstances of their birth. An absence of given values for the constants and a lack of appreciation for social behavior and reproductive behavior are large stumbling blocks to the application of predictive mathematical models to natural populations.

Notes

1. M.G. Levy, M. Tunis and H. Isseroff, 1973. 'Population control in snails by natural inhibitors'. *Nature* 241: pp.65-6.
2. S.M. Rose and F.C. Rose, 1961. In 'Mechanisms in Biological Competition'. *Symp. Soc. Exper. Biol.* 15: p.207, Cambridge University Press, Cambridge.
3. L.M. Roth, 1943. 'Studies on the gaseous secretion of *Tribolium confusum* Duval. II. The odiferous glands of Tribolium confusum'. *Ann. Ent. Soc. Amer.* 36: pp.397-424.
4. _____ and R.B. Howard, 1941. 'Studies on the gaseous secretion of *Tribolium confusum* Duval. I. Abnormalities produced by exposure to a secretion given off by the adults'. *Ann. Ent. Soc. Amer.* 34: pp.151-72.
5. P. Alexander and D.H.R. Barton, 1943. 'The excretion of ethylquinone by the flour beetle'. *Biochem. J.* 37: pp.463-5.
6. S. Pickering, 1917. 'The effect of one plant on another'. *Ann. Bot.* 31: pp.181-7.
7. E.F. Davis, 1928. 'The toxic principle of *Juglans nigra* as identified with synthetic juglone, and its toxic effects on tomato and alfalfa plants'. *Amer. J. Bot.* 15: p.620, abs.
8. A.B. Massey, 1925. 'Antagonism of the walnuts (*Juglans nigra* L. and *J. cinercc* L.) in certain plant associations'. *Phytopathology* 15: pp.773-84.
9. L. Overland, 1966. 'The role of allelopathic substances in the "smother crop" barley'. *Amer. J. Bot.* 53: pp.423-32.
10. E.L. Proebsting, 1950. 'A case history of a "peach replant" situation'. *Proc. Amer. Soc. Hort. Sci.* 56: pp.46-8.
11. L.J. Webb, J.G. Tracey and K.P. Haydock, 1967. 'A factor toxic to seedlings of the same species associated with living roots of the non-gregarious subtropical rain forest tree *Grevillea robusta*'. *J. Appl. Ecology* 4: pp.13-25.
12. H.L. Ratcliffe, 1961. 'Discussion of the papers by Christian and Snyder'. *Proc. Nat'l. Acad. Sci.* 47: pp.455-8.
13. M. Williamson, 1972. *The Analysis of Biological Populations.* Edward Arnold, London; Crane, Rerssak, New York.
14. R. Pearl, 1930. *Introduction to Medical Biometry and Statistics.* Saunders, Philadelphia.
15. P.F. Verhulst, 1938. 'Notice sur la loi que la population suit dans son accroissement'. *Corresp. Math. Phys.* 10: pp.113-21.
16. R. Pearl, and L.J. Reed, 1920. 'On the rate of growth of the population of the United States since 1790 and its mathematical representation'. *Proc. Nat. Acad. Sci. U.S.* 6: pp.275-88.
17. R.L. Snyder, 1968. 'Reproduction and population pressures'. In *Progress in Physiological Psychology,* eds. E. Stellar and J.M. Sprague, Vol. 2: pp.119-60, Academic Press, New York.
18. F.A. Crew and L. Mirskaia, 1931. 'Effect of density on adult mouse populations'. *Biologia Generalis* 7: pp.239-50.
19. J.R. Clark, 1955. 'Influence of numbers on reproduction and survival in two experimental vole populations'. *Proc. Royal Soc. (Lond.)* B 144: pp.68-85.
20. V.C. Wynne-Edwards, 1965. 'Self-regulating systems in populations of animals'. *Science* 147: pp.1543-8.
21. J.A. Wiens, 1966. 'On group selection and Wynne-Edward's hypothesis'. *Amer. Scientist* 54: pp.273-87.
22. C.R. Terman, 1965. 'A study of population growth and control exhibited in the laboratory by prairie deermice'. *Ecology* 46: pp.890-95.

6 NATURAL REGULATION OF POPULATIONS

In the first chapter of this book I noted the existence of opposing schools of thought about control of population growth. Extrinsic agents such as food supply, weather, predators, and disease might be responsible for determining population size on certain occasions. However, studies of the actions of each of these agents on population dynamics leave reasonable doubts about their effectiveness as sole regulatory factors. In many instances populations could exist at higher levels of density if food were utilized to capacity. Weather may be too variable in its actions to have much effect on abundance of animals over a long period of time. Also adaptive evolution has produced animals that are relatively immune to even the harshest climatic conditions. Predators are effective at times, but again structural and behavioral adaptations of the prey are constantly negating the impact of predation on population density. Furthermore, predators themselves may possess intrinsic mechanisms for population control. Finally, parasites and pathogenic organisms do not exist to limit numbers of their hosts. As a matter of fact, the strategy of parasitic organisms is best served if the host survives.

For several cogent reasons many ecologists believe that inherent behavioral and organismal characteristics of the members of the population themselves provide the intrinsic mechanisms that regulate population growth. This belief requires an added condition — that regulation of the population is an adaptive response which confers an advantage to the members collectively. Then, intrinsic mechanisms of population control, having adaptive advantage, would be subject to natural selection and selective pressures would act on groups of animals or whole populations, rather than on individual organisms. Such evolutionary thought is severely criticized by many.

The habitat contains only a finite total quantity of nutrient elements such as nitrogen, phosphorous, potassium, calcium, iron, and iodine. The continuity of life depends upon the repeated circulation of these essential elements. The maximum biomass (with respect to a single species) that could be maintained in a given region could be calculated if all variables were known. Ideally, the maximum numbers of herbivores that could be maintained in a grassland community would depend on the energy cycle of the community, provided the herbivores were optimally spaced and possessed no behavioral patterns that would disrupt the spacing. The herbivores in this ideal community would eat a maximum sustained yield of vegetation without interfering with the plants' capacity to reproduce and maintain themselves. The strategy in this case would be something akin to the concept of 'maximum sustained yield' used by wildlife managers and agronomists. It is doubtful if such an arrangement exists anywhere in nature. Furthermore, to attribute such strategy to organisms with limited faculties for reasoning and communication would be carrying the analogy to absurdity.

Another line of reasoning explores the inherent capacity of increase possessed by all living organisms, large and small. Reproductive potential for any species is so great that populations would easily grow beyond the carrying capacity of the environment if unchecked by intrinsic or extrinsic forces. The proponents of intrinsic regulation of population have no faith in the strength of extrinsic agents to restrict growth of populations. But at the same time, they visualize the destruction of the habitat and its capacity to generate nutrients as a catastrophic result of unlimited population growth. Without adaptive intrinsic regulatory mechanisms populations would soon outstrip their food supply and create conditions that would lead ultimately to extinction of the species.

Members of the school of intrinsic regulation accept the proposition that food is the ultimate limiting factor, the critical resource. Of course there are other requirements to be satisfied — nest sites, shelter, water, singing grounds, etc., but food, dependent on energy flow, is produced in limited quantities and the competition for it is fierce.

Many theories for intrinsic or 'natural' regulation of population growth have been proposed, but all have one tenet in common — density-dependency. The mechanisms also must work through the three population forces that determine abundance: reproduction, mortality, and dispersal. It is interesting that certain theories envision the mechanisms tied to reproduction, others to mortality and still others to dispersal. A few theories propose mechanisms tied to all three forces. It would be difficult to arrange the theories into inclusive categories, for there are overlapping components. Generally, the mechanisms envisioned are tied to social interaction (behavioral characteristics) or genetics.

The Social Stress Theory

Christian's social stress theory, which embodies the principles of psychological stimuli, received a great deal of attention when it was presented in 1950.[1] In its original form this theory was proposed to explain the cyclic nature of certain arctic and subarctic populations of

Figure 25. Changes in the Abundance of the Lynx and the Snowshoe Hare as Indicated by the Number of Pelts Received by the Hudsons Bay Company (Redrawn from paper by MacLulich, D.A. 1937. Fluctuations in the numbers of the varying hare *(Lepus americanus). Univ. Toronto Studies, Biol. Ser.,* No.43.)

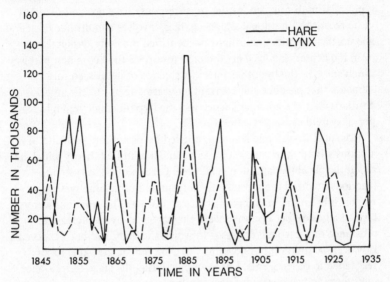

mammals (Fig. 25) and utilized the classic stress concepts of Professor Hans Selye of Canada.[2,3]

William Cannon was the first to develop the idea that organisms react consistently to unfavorable situations in terms of highly integrated metabolic activities.[4] These reactions follow a stereotyped sequence which Selye designated as (a) alarm, in which the organism goes into a state of shock with falling temperature, irregular blood sugar levels, and depression of nervous functions; (b) defense, during which the organism tends to reverse the changes that occurred during the initial alarm reaction, develops an increasing resistance to the stressor, and adapts itself to a new situation; and (c) exhaustion, when the organism loses the adaptation acquired during the preceding stage.

Experimentally these reactions may be produced in various ways: by forced immobilization, abdominal laparotomy, intraperitoneal injection of histamine, insulin, and formalin, etc. Almost any stimulus — crowding, isolation, or even muscular exercise — may become a stressor. Some of the noticeable manifestations of the response to nonspecific stress include increase in the concentration of adrenal steroids in the blood, depletion of neurosecretory granules in the neurohypophysis,[5] enlargement and hyperactivity of the adrenal cortex, eosinopenia, lymphopenia, thymolymphatic involution, and gastrointestinal ulcers.

A number of details concerning the pathways by which different types of stimuli bring about the stress response have been worked out more recently. Emotional stress, e.g. fear, involves a multitude of neural and cortical circuits that cannot be identified precisely. Projections from the centers involved are directed towards a final common pathway which starts in the hypothalamus.[5] It consists of neurosecretory elements that produce and transmit a releasing factor to the anterior pituitary via the median eminence and the hypothalamic hypophyseal portal system causing release of ACTH.

Nonspecific systemic reactions elicited by prolonged exposure to noxious stimuli were studied by Selye under a unified theory which he called the general-adaptation syndrome. Certain of these systemic reactions involved the reproductive organs and others the systems responsible for maintaining resistance to disease. The reactions were termed adaptive because they were characterized by an increased resistance to a particular stressor to which the body had been exposed. However, adaptation to one agent was acquired 'at the expense' of resistance to other agents. For example, during the stage of resistance, reproduction becomes subservient to functions of more immediate concern for survival of the organism. Also, during the stage of resistance an organism might become more liable to succumb to infectious or parasitic diseases.

If exposure to abnormal conditions continues, adaptation 'wears out' and many lesions characteristic of the alarm reaction (AR), e.g. involution of the thymico-lymphatic system, loss of adrenal lipids, gastrointestinal ulcers, reappear and further resistance becomes impossible. The stage of exhaustion represents the sum of all nonspecific systemic reactions that ultimately develop as a result of prolonged over-exposure to noxious stimuli. Exhaustion results in death.

Involution of the testis and the male accessory sex organs is a characteristic effect of exposure to systemic stress. The mature spermatogenic cells are particularly sensitive, but in the case of intense and prolonged exposure, all elements of the seminiferous tubules, except the Sertoli cells and the most immature elements of the spermatogenic series, disintegrate. The Leydig cells also involute and the testis assumes an almost embryonic appearance. Involution of the testis and accessory sex organs occurs simultaneously with other manifestations of the AR, such as thymic involution and adrenocortical enlargement. Emotional stress is capable of producing these characteristic changes in the testis.

The characteristic response of the female sex organs to systemic stress is ovarian atrophy accompanied by more or less permanent anestrus. Upon continued exposure to stressors, the ovary returns to normal and vaginal cycles reappear.

Decreased resistance to infectious and parasitic diseases and the occurrence of degenerative diseases in stressed animals are related chiefly to adrenal physiology, although there are other organ systems involved as well. It is an established fact that adrenalectomized animals have very little ability to adjust to stressful stimuli, such as extremes of temperature, trauma, bleeding, infections, toxins, etc. Adrenal exhaustion was once thought to mimic adrenalectomy, but true adrenal exhaustion has not been demonstrated *in vivo*. Whatever effects occur are more likely due to the direct effects of certain adrenal corticoids secreted during stress. For example, large amounts of anti-inflammatory adrenal steroids are secreted in response to pituitary ACTH. This response is useful in enabling an organism to survive during an emergency by suppressing energy depleting inflammatory reactions. On the other hand, it may be harmful inasmuch as it permits the spread of infections.

Other stress responses are equally responsible for lowered resistance to disease. Lymphopenia, thymicolymphatic involution, and macrophage reduction diminish an organism's ability to counteract infectious agents because of the reduction in chronic inflammatory cells, antibody production, and phagocytic activity. Even the production of granulation tissue, which walls off infectious agents with a fibrous capsule, is diminished during stress.

The degenerative diseases that are said to occur because of prolonged stress may not be very important in natural populations. Thus far, such pathological changes as hypertension, arthritis, arteriosclerosis, nephrosclerosis, and gastrointestinal ulcers have been produced by administering pituitary and adrenal hormones to experimental animals sensitized by unilateral nephrectomy and by a high dietary load of sodium chloride. There are, of course, exceptions to any general statement. Membranous glomerulonephritis is an extremely common disease in most rodent species and may be an important factor in stressed animals at high population densities. This disease was responsible for many deaths among woodchucks studies in south-central Pennsylvania.[6]

Christian proposed exhaustion of the adrenopituitary system

subsequent to the 'stresses' inherent in high population levels, severe climatic conditions and the demands of the spring breeding season as the principle cause of cyclic declines in mammalian populations. He listed nine stresses that were likely to occur at peak densities: (a) food scarcity; (b) lack of proper cover; (c) increased muscular exertion resulting from longer food forage trips; (d) fights with other individuals; (e) increased exposure to cold from longer forage trips and inadequate cover; (f) fighting resulting from territorial encroachment; (g) utilization of inadequate foods; (h) increased exposure to predators due to lack of cover as well as migration of predators into areas of abundant food supply in the form of peak population; and (i) nutritional deficiencies. The increasing length of daylight in the spring further stimulating the pituitary and the energy demands of reproduction were postulated as the final insults to an already overtaxed endocrine system.

As first proposed, the stress theory included many extrinsic elements (e.g. food scarcity, climate, and predation), but tied these with the endocrine system to explain their action in regulating population growth. The more important stressors were believed to emanate from intraspecific competition for food, cover, and space. Animals might normally be able to withstand severe climate episodes without undue losses at low population density when food and cover were not in short supply. But the same conditions would increase mortality significantly when animals were already in a state of stress due to intense social interaction at peak densities.

Selye's general adaptation syndrome provides a set of intrinsic mechanisms that would affect reproductive and mortality rates in a density-dependent manner. Population pressures affect various stages of reproduction depending on their intensity. The same stressors lower the general resistance of animals to infectious and parasitic diseases. Incidentally, Professor Selye designated the stimulating agents as stressors and the systemic response as the state of stress. The social stress theory fits the observed elements of cyclic population phenomenon. The rapid build-up from low population levels has been observed in both experimental and natural populations. The growth phase is accompanied by low interaction rate, high reproductive performance and relatively low mortality especially in the absence of predator pressure. Inhibition of reproduction has been observed in various forms at peak densities. Finally, the 'crash' usually occurs rapidly with the preponderance of deaths caused by epizootics of disease. The infectious agents responsible for the outbreaks of disease have ranged widely from one geographic location to another and from one time to another. The pathogen seems to be an opportunist. Whatever germ happens to be prevalent when the stage of stress reaches the critical point is the cause of the epizootic.

Christian utilized a number of caged populations of mice to obtain experimental proof for his theory. Adrenal enlargement and gonadal atrophy occurred when male mice were grouped; females generally responded only when males were present. Inhibition of reproduction and enhancement of mortality at asymptotic population levels in freely growing populations of mice were also associated with adrenal enlargement and other characteristic stress-induced responses. Christian then presented a revised theory that population growth and decline were regulated by series of neuroendocrine feedback mechanisms activated by socio-psychological factors. The system responds to changes in the number of animals in such a way that population growth is self-limiting and self-regulating.

Christian's later papers on the social stress theory seldom mentioned cyclic phenomenon. Instead, the theory became an all-inclusive general population law to explain most, if not all, of the phenomena of population dynamics. Critics of the social stress theory question the application of data from caged populations because of the unnatural densities involved — approaching one million per acre in some experiments. Also, one field study in particular is cited as showing that rate of social contact remains constant as density increases.[7] Antagonists also point to studies of mouse populations that show no relationship between density and weights of the adrenal glands.

In fairness to theorists and critics alike, it would seem that until recently the debate has raged with insufficient evidence to make a final judgment for either side. Critics probably underestimated the sensitivity of neural centers perceiving extrinsic stimuli. Neural-endocrine-gonadal systems that respond to such nebulous stimuli as variations in light intensity and odors should certainly react to emotional stimuli emanating from social interaction. As a matter of fact, John Mason and his colleagues at the Walter Reed Army Institute of Research have demonstrated that psychological stimuli are extraordinarily potent in eliciting the stress response.

Field studies attempting to show an adrenal relationship to population density have been largely unsuccessful, but such efforts were probably doomed to failure from the start. Adrenal weight is the product of several factors that bear no relationship to population density, i.e., season, sexual cycle, age, and nutritional state, to name the most obvious. In laboratory cages such environmental variables were eliminated or equalized to make the one parameter, population density, measurable. To say that under natural conditions adrenals do not respond to population pressures because adrenal weights show no relationship to population density in field studies is illogical. After all, data from laboratory studies have been offered to support many theoretical conditions about population phenomena. The complexity of the natural system is supposedly the reason for conducting laboratory experiments in the first place. Thus it seems incongruous that critics would expect to find a clear-cut example of adrenal-density interplay in nature.

A few field studies demonstrating unequivocal evidence of stress response affecting reproductive physiology were reviewed in Chapter 2. Total litter resorption in swamp rabbits following sudden crowding and inhibition of sexual maturity in cyclic voles during periods of high density were perhaps the best examples, since food shortage was not a confounding factor. In a great many cases, poor reproduction and high mortality from disease and parasites are attributed simply to nutritional deficiencies. The obvious density-dependent relationship is too often ignored. It cannot be argued that deer do not starve to death or that reproduction is not affected by shortage of browse, but neither can it be denied that stress does not play a role, either through increased intra-specific competition for a diminishing resource or through the powerful psychological stimuli associated with hunger.

Wynne-Edwards' Social Convention Theory

Wynne-Edwards[8,9] visualizes the evolution of a diverse array of behavioral patterns that regulate population density. Dominance-subordination hierarchies, territoriality, and other social conventions are considered evolved mechanisms that relay information concerning numbers, especially numbers in relation to available resources, to the individuals of a population. Various psychological and physiological mechanisms then act to adjust population through their effects on reproduction, mortality, and movement. Wynne-Edwards has incorporated many of the elements of the social stress theory into this theory, so much in fact that, in my opinion, the two are nearly identical. His theory is more an elaboration of the theory of density-dependent endocrine mechanisms developed by Christian[10] with more details concerning the way these mechanisms might work in nature.

Wynne-Edwards suggests that animals become tuned to population density through various kinds of group-oriented displays, such as flocking movements, territorial singing and drumming, etc. He does not imply that animals are consciously aware of population density or its significance in terms of resource management, but only that individuals automatically exhibit a physiological response to contacts or to intensity of social interaction.

Wynne-Edwards believes that several artificial and harmless types of competition have evolved as buffer systems in many species to stop population growth at a level below that imposed by food exhaustion. The territories that birds defend so fiercely during the breeding season are envisioned as part of a system that distributes food resources according to the productivity of the habitat. Territoriality excludes a portion of the population from breeding. Thus, population growth is limited and the food supply will not be exhausted. Presumably territories would be smaller in more productive habitats, which would allow a greater population increase because more individuals would produce offspring.

Genetic Feedback Mechanism

The British geneticist E.B. Ford[11] was one of the first to point out the possible importance of genotypic changes in population regulation. He suggested that selective pressures were relaxed during the growth phase, with the result that genetic variability increased and many inferior genotypes survived. When less propitious conditions returned, these inferior individuals would be eliminated through increased selective pressures, causing the population to decline and at the same time reducing genotypic variability.

The concept of a genetic feedback mechanism functioning to regulate populations of plants, herbivores, predators, and parasites is supported by mathematical analysis of data on population dynamics[12] and studies of natural populations.[13,14] In this case, excessive population density, i.e. in excess of the carrying capacity of the habitat, is considered the selective pressure in an interacting system. All levels of the food cycle are affected with the result that new genotypes predominate in the plants and animals of the community which adjust population densities to fit the carrying capacity of the environment. The ultimate end is stability of the community with regard to numbers of different plants and animals. Periodic disturbances of the ecosystem such as catastrophic climatic events, forest fires, increased salinity, or the introduction of a new parasite cause pronounced fluctuations in numbers of plants and animals and may exert a selective force for different genotypes. The feedback part of this system involves action and reaction on the part of interacting populations. When a new parasite is introduced to a community, the first result may be a tremendous increase in the population of the parasite and a corresponding decrease in the population of the susceptible host, but eventually the resistant members of the host population, with a new genotype predominating, will reproduce and accommodation with the new parasite will be achieved. Characteristic levels of density for the host animals in the community would not necessarily be the same as before.

Few ecologists would argue that changes in genotypes would not be effective in influencing average abundance of a species on a long-term basis. Whether such variations operate during short-term oscillations of cyclic species to change reproductive and mortality rates is debatable.

Data from studies of *Microtus* show different gene frequencies in emigrating individuals and in populations at different densities.[15] The genes measured have nothing to do with reproductive performance or with quality of the individuals with respect to resistance or susceptibility to disease. However, such studies demonstrate that significant changes in gene frequencies can occur over a relatively short period of time. If selective pressures act rapidly to sort out genotypes related to reproductive efficiency or resistance to disease there could conceivably be a density-dependent change in reproductive and mortality rates. Reproductive and immunological mechanisms, to name two rather complex systems related to population productivity, would be under the control of a wide assortment of allelic genes. It remains to be seen whether selective pressures within the population are capable of changing genotypic frequencies sufficiently and rapidly enough to account for short-term density-dependent responses.

Natural Selection of Mechanisms for Population Control

Adherents of natural population control are almost forced to accept the idea that regulatory mechanisms are adaptive and thus subject to natural selection. Critics of biotic adaptation are inclined to force the choice, since they are usually also critical of the intrinsic school of population regulation.

The natural selection of alternative alleles in a Mendelian population is called *genic selection*.[16] The natural selection of more inclusive entities, a herd, a pack, a colony or a population for example, is called group selection, a term introduced by Wynne-Edwards. Genetic diversity among individual organisms plus genic selection produces organic evolution with organic adaptation and biotic evolution without biotic adaptation.[16] Random alterations of the sizes of populations, especially extinction, also produces biotic evolution without biotic adaptation.

Genetic diversity among populations plus the reduction or extinction of the less fit population (group selection) produces biotic evolution with biotic adaptation.

An organic adaptation is a mechanism 'designed' to promote the success of an individual organism, as measured by the extent to which it contributes genes to later generations of the population. Biotic evolution is any change in biota. A biotic adaptation is a mechanism 'designed' to promote the success of a biota, as measured by the lapse of time to extinction. Evolutionary design, of course, happens by accident or chance and is sorted out by selective pressures.

It is true that the fact that a population survives through a succession of generations is not evidence for the existence of biotic adaptation. The survival of the population may be merely an incidental consequence of the organic adaptations by which each organism attempts to survive and reproduce itself. To determine whether one population is more successful than another is indeed a difficult exercise in judgment, but this is probably not what most ecologists that favor intrinsic population regulation are chiefly concerned with.

I doubt very much whether any population biologist would deny that genic selection is important nor even that it is the most powerful force for evolutionary change. The general adaptation syndrome visualized by Selye presumes intrinsic adaptive mechanisms that allow individual organisms to cope with the exigencies of life. Such mechanisms should theoretically increase an individual's inclusive fitness and thus maximize its ability to contribute genes to later generations. Stress inhibits reproduction when the environment is especially harsh but such inhibition is normally of short duration. Reproduction is sacrificed to insure survival and the chance to reproduce at a later time. In this respect Terman has demonstrated that female *Peromyscus* reproductively inhibited in dense populations reproduce well when moved to less crowded environs.[17] It may be fortuitous that inhibition of reproduction occurs at high population densities but such an occurrence does slow population growth. In this case what is good for the individual female happens to be good for the whole population. Selye was concerned only with individual adaptation, but Christian, for one, considered these intrinsic mechanisms that exist in individuals as important adaptive mechanisms for the population. They work to insure population survival only because these mechanisms are activated by psychological stimuli that depend on population pressures. In brief these intrinsic mechanisms likely evolved in individuals but the selective forces may have been largely emanating from population pressures. The primary value lies in the way the system works. If it works to insure the survival of the population at the expense of individuals the adaptation must be considered biotic according to definition.

Freely-growing caged populations of brown house mice reach a point at which there is tremendous wastage of flesh and with it a reduction in the total gene pool. Conceivably, if food were an ultimate limiting factor the sizeable losses would assure survival of part of the population. If all members of the population were equally competitive for the available nutrients and if even a portion of the female members continued to reproduce, the results would most likely be population extinction. The expectation of extinction must be qualified because there are no examples of populations where these conditions have existed. Also, confined populations with unlimited food supplies become extinct, so there is no precedence for making deductions. Nevertheless, the proponents of intrinsic population regulation believe that poorly-adapted populations do become extinct. The vacuum created is filled by emigrants from successful populations. Presumably the emigrants possess the intrinsic mechanisms that react to population pressures. In this way population regulation is achieved, even though the intrinsic mechanisms originally evolved through genic selection.

On the face of it group selection would play a minor role and I would not quarrel too much with the interpretation. The evolution of territoriality, dominance hierarchies and social conventions through group selection is another matter entirely. In most instances such phenomena can be explained adequately on the basis of individual fitness. Territoriality and dominance behavior, for example, would maximize the individual's contribution to the gene pool. The argument again revolves around the possibility that such phenomena would favour survival of the group or population. I believe in most cases an unbiased observer would have to admit that such was the case. Whether these behavioral patterns can be called biotic adaptations would depend again on the strength of group selection. Natural selection operates on each individual as a unit and presumably on each group or population as a unit. Since the ratio of individuals to groups is so great, evolution would proceed much more rapidly on an individual basis. Moreover, the individual's lifespan is only a brief second when compared to the span of existence of a group or population, which contributes still greater numbers of individual units for selection.

Reasoning and interpretation of natural phenomena will never substitute for positive proof, but they will have to suffice, where evolution is concerned. In Chapter 1 it was pointed out that one part of a protein chain making up an antibody required two separate genes for its synthesis, which is contrary to the prevailing one gene — one enzyme (protein) theory. Genic selection is characterized by the natural selection of alternative alleles, but the protein chain of the antibody would require the selection of two sets of alleles. The situation with respect to antibody production may be unusual in terms of genic selection, but it lends credence to the thought that occasionally two units are selected to accomplish a result that neither one alone would be capable of achieving.

Altruism and Natural Selection

Altruism is defined as any self-sacrificing behavior that benefits another individual. Co-operative interactions when they involve closely related individuals would tend to maximize the representation of an individual's own germ plasma in the population. A male that exhibits solicitous parental care and protection to its offspring would have this effect. Benevolent self-sacrifice among genetically different individuals, however, would tend to preserve the gene pool of the group. In either case the genes controlling such behavior would be subject to natural selection.

Co-operative interaction between closely related individuals would normally preserve the gene or genes responsible for the advantageous behavior. Self-sacrifice or foregoing reproduction would diminish the individual's contribution to the gene pool, but because the recipient individuals possess similar germ plasma the ultimate result in terms of genes passed on to future generations might actually be enhanced.

I would reason that altruistic behavior among unrelated individuals might have the same result. Consider an animal that possesses a genetic program for benevolent behavior and suppose that such behavior increases the probability of death. The genes responsible for this pattern of behavior would soon be eliminated from the gene pool even though other individuals might survive a while longer. For altruistic behavior to have any long term value and for the genes to be selected would require at least two animals to have the prerequisite behavior. Benevolent behavior, even if it increases the probability of death, would be acceptable if other individuals in the group returned the favor and if such behavior was required only under the most adverse circumstances. In other words, altruism is not a behavior that is bandied about. Common sense would dictate that continuous self-sacrificing behavior would be self-defeating and would surely not evolve under natural selection.

Actually the term altruistic pertains only to benevolent self-sacrifice among genetically different individuals.[16] George C. Williams maintains that most examples of co-operation and self-sacrifice in a group project (e.g. social insects) are confined to genetically similar groups. 'Whenever there are behavioral mechanisms by which parents aid their offspring, there will inevitably be times when such aid is provided "by mistake" to unrelated individuals.' This biologist believes that altruism is not an example of biotic adaptation.

The woodchuck *(Marmota monax)* lives a solitary existence in the north-eastern United States, southern Canada, and south-eastern Alaska but its closest relatives, the marmots *(M. flaviventris, m. caligata,* etc.) generally take up the colonial existence. I mentioned in Chapter 4 that woodchucks are more agonistic during the spring breeding season and again when the young of the year are dispersing from their maternal dens. Oddly enough I observed what might be considered the roots of colonial behavior in woodchucks during the spring breeding season each year. Pairing of woodchucks occurs consistently only during late winter and early spring. A pair in this case was two animals that occupied a single burrow system without any outward sign of antagonism toward one another. On ten occasions the two animals involved were shot after they had been observed long enough to establish their behavior toward one another. Eight of these pairs included a male and a female. But one pair consisted of a large, sexually mature adult male and a smaller immature yearling male and another consisted of a pregnant female and a nulliparous yearling female. In most instances the pairs were observed for close to an hour outside their burrow entrances before an attempt was made to collect the animals. The homosexual pairs were behaving contrary to the norm but whether the behavior was altruistic is open to question. Whatever the explanation, misplaced parental behavior or something else, the animals were probably not closely related since dispersion of the young had occurred during the preceding year. Nevertheless the existence of homosexual pairs among otherwise solitary animals might provide the roots of colonial behavior. The marmots generally live in colonies in mountainous regions but exist as solitary animals where forage is plentiful. Adaptability may be the key; behavioral patterns seem to develop as the environment dictates. In this instance co-operative interaction would be advantageous to the group. It is conceivable that pairs of animals would be selected because they possessed the necessary genes to reduce agonistic behavior during the breeding season, which would be the first step toward developing true colonial behavior.

Summary and Conclusions

Density-independent factors, climate and food supply especially, affect reproduction, mortality, and movement to the extent that these can be considered regulatory on occasion. Such factors set the limits of abundance for a population in its physical habitat. Unfavorable climatic factors are nearly always density-independent; however, food restriction may be considered density-dependent if competition is increased. The pressure of food restriction may set off a whole series of responses that ultimately have the result of reducing density overall and redistributing the members of the population geographically. It is doubtful whether the population is very often limited because some animals starve to death.

Density-dependent factors operating primarily through competition set the limits of abundance of a population in its social environment. The social stress theory emphasizes the role of endocrine feedback mechanisms responding to the increased social pressures in growing populations. Other mechanisms involving psychological and physiological responses to social interactions may be equally important as regulatory mechanisms since reproduction and, in some instances, mortality are affected. These mechanisms fall into two broad categories: (a) those related to early experience; and (b) specific response to certain kinds of social contacts.

Intrauterine effects that cause permanent aberrations in behavior patterns of the offspring and hormonal stimulation during the neonatal period are included in the first category. Alterations of the hormonal status of newborn animals have lasting effects on biological functions. Population pressures affecting the gonadal and adrenal systems in the neonatal period could result in limited reproductive potential in adult life. These effects combined with traits or behavioral patterns acquired while growing up in crowded environs might easily produce a preponderance of subordinate animals with poor reproductive performance. The existence of these or similar mechanisms might explain the social structures that become established in laboratory populations of mice and account for the characteristic sigmoid growth form as well. Animals that attain the dominant status early in life would be expected to have a competitive advantage throughout life. Under these circumstances such animals would be little affected by increasing population pressures. In contrast, animals born during periods of high population density, when presumably the essential prerequisites of life are at a premium, would assume a subordinate position. These animals might be forced to move to the least desirable parts of the habitat, a compensatory response in itself, or under dire circumstances increased mortality among this class of animals combined with poor reproduction would decrease population density accordingly. A few persistent dominant animals would assure a nucleus from which subsequent population growth would spring when the essential resources of the habitat were again abundant. Whatever the psychological and physiological mechanisms, this is the scheme I prefer to explain natural regulation of population density.

The effects of early experience in producing a dominance hierarchy would negate the natural selection of either an optimum or a flexible reproductive rate. Reproduction of dominant animals would be maximized and that of subordinates minimized. The productive potential, genetically determined, would have been the same in both social groups; however, the potential would be realized only with the achievement of the dominant status. In fact, reproductive rate is a population attribute or collective statistic. The inverse relationship between reproduction and density in populations of brown house mice (Fig. 21) and prairie deer mice is a case in point. When population density was very high, a majority of the females simply did not reproduce. Reproductive rate decreased, but in this case, one class of animals, the dominants, reproduced as well as before. The reproductive rate of the dominant females was not related to population density at all. The relationship between reproduction and population density existed solely as a population attribute.

Disruption of established estrous rhythm accomplished by grouping female mice and male-induced synchrony of estrus (Chapter 2) are examples of physiological responses to specific kinds of social interaction. The pregnancy block induced by exposure to an alien male appears to be the only one of the so-called specific responses that might play a role in population regulation.

Genetics is undeniably important in the intrinsic scheme of population regulation because reproductive potential is an inherited trait. However, I have suggested an alternative to the natural selection of an optimum reproductive rate[12] or a maximum reproductive rate.[16] The consensus is that natural selection will tend to maximize potential rate of increase of a species for the environment in which it lives. However, if the environment is in a state of flux it is difficult to see how this might be achieved. One can argue just as well that natural selection would tend to favor a mechanism for a flexible reproductive rate, a maximum response when the environmental conditions were favorable and a minimal response when the reverse were true. The scheme outlined in the preceding paragraphs would fit this description. In this instance, the selective pressure is an unstable environment that imposes variable conditions on the members of a population.

The population is itself an environmental factor for the individual and, as such, may often constitute a selective force of considerable magnitude. But it should be kept in mind that many populations undergo pronounced changes in density within relatively short periods of time. Here again, the population, as a selective force, has rather unstable characteristics. Since it is adaptive for individuals to react efficiently in accord with other members of the population, one can argue further that mechanisms allowing a flexible response with respect to reproduction and density are advantageous to the individual. At high densities reproduction would be sacrificed but in return for a better chance to survive for the individual. Unrestricted reproduction at low densities would maximize the number of progeny having a genotype that allows a flexible reproductive rate in response to population density.

In general the 'strategy' of evolution is to expand and preserve the gene pool. In a sense the gene pool is the total population, thus it was inevitable that ecologists would symbolize their rhetoric somewhat as follows. In the long run the individual doesn't matter, it is the survival of the species that is of paramount importance. No wonder then that arguments about individual versus group selection get caught up in semantics. We say that the 'strategy' of evolution is to maximize an individual's contribution of the germ plasma to the population and in the same breath repeat the general laws of speciation.

In general it is adaptability that counts. Individuals that have the capacity to change when necessary are more likely to survive. And populations of animals that have the capacity for adaptability built into their gene pools are also more likely to survive. Individual adaptability is epitomized in the range of defensive antibodies that are available to the individual organism, which is estimated to number into the thousands.

A population must have an even greater number of defensive (adaptive) responses at its disposal because of the immensity of the gene pool. The compensatory response in population dynamics will be discussed in the next chapter.

Notes

1. J.J. Christian, 1950. 'The adreno-pituitary system and population cycles in mammals'. *J. Mamm.* 31: pp.247-59.
2. H. Selye, 1936. 'A syndrome produced by diverse noxious agents'. *Nature* 138: p.32.
3. —— 1946. 'The general adaptation syndrome and the diseases of adaptation'. *J. Clin. Endocrinology* 6: pp.117-230.
4. W.B. Cannon, 1914. 'The emergency function of the adrenal medulla in pain and the major emotions'. *Am. J. Physiol.* 33: pp.356-72.
5. E. Scharrer, and B. Scharrer, 1963. *Neuroendocrinology.* Columbia University Press, New York and London.
6. R.L. Snyder, D.E. Davis and J.J. Christian, 1961. 'Seasonal changes in the weights of woodchucks'. *J. Mamm.* 42: pp.297-312.
7. O.P. Pearson, 1960. 'Habits of *Microtus californicus* revealed by automatic photographic records.' *Ecol. Monog.* 30: pp.231-49.
8. V.C. Wynne-Edwards, 1962. *Animal Dispersion in Relation to Social Behavior.* Hafner, New York.
9. —— 1965. 'Self-regulating systems in populations of animals'. *Science* 147: pp.1543-8.
10. J.J. Christian, J.A. Lloyd and D.E. Davis, 1965. 'The role of endocrines in the self-regulation of mammalian populations'. *Recent. Progr. Horm. Res.* 21: pp.501-78.
11. E.B. Ford, 1931. *Mendelism and Evolution.* Methuen, London.
12. D. Pimentel, 1961. 'Animal population regulation by the enetic feedback mechanism'. *Amer. Naturalis.* 95: pp.65-79.
13. D. Chitty, 1957. 'Self-regulation of numbers through changes in viability.' *Cold Springs Harbor Symposia on Quantitative Biology* 22: pp.277-80.
14. —— 1960. 'Population processes in the vole and their relevance to general theory'. *Canadian J. Zool.* 38: pp.99-113.
15. C.T. Krebs, M.S. Gaines, B.L. Keller, T.H. Myers and R.H. Tamarin, 1973. 'Population cycles in small rodents'. *Science* 179: pp.35-41.
16. G.C. Williams, 1966. *Adaptation and Natural Selection, A Critique of Some Current Evolutionary Thought.* Princeton University Press, Princeton, New Jersey.
17. C.R. Terman, 1973. 'Recovery of reproductive function by prairie deermice *(Peromyscus maniculatus bairdii)* from asymptotic populations'. *Anim. Behav.* 21: pp.443-8.

7 COMPENSATORY MECHANISMS

'I have often thought,' wrote Justus Liebig to his friend Duclaux, 'in my long and practical career and at my age (69 years) how much pains and how many researches are necessary to probe to the depths a rather complicated phenomenon. The greatest difficulty comes from the fact that we are too much accustomed to attribute to a single cause that which is the product of several, and the majority of our controversies come from that.' Over-simplification is an error often committed by scientists in their drive to discover basic principles that relate diverse facts. If the terms are general enough to incorporate complex phenomena, they are likely to be hazy and ambiguous. Liebig (1803-73) was a German chemist interested in agriculture and population problems. The last two sentences were selected verbatim from *Principles of Animal Ecology*, a truly remarkable book on ecology written by W.C. Allee, Alfred E. Emerson, Orlando Park, Thomas Park, and Karl P. Schmidt and published in 1949. The whole problem of population growth was treated by these ecologists as extremely complex.

Natural regulation of population and control of numbers by intrinsic mechanisms existing as neurological, endocrinological and physiological components of member organisms were apparently not envisioned by Allee and his associates. They spoke of population control in social insects, the elimination of drones in honeybee populations at the approach of winter for example, but this is as close as they came to a concept of natural regulation of population density. The first theories concerned with natural regulation were apparently the proposal set forth by Christian linking the 'adreno-pituitary system' to population cycles in mammals and John Calhoun's[1] concept of self-limitation of populations. Both ideas were published in 1950.

Allee and his associates did, however, present a schematic representation of the interplay of factors that affect populations called 'population integration'. Integration was considered the interaction of pressures caused by ecological and genetic factors. These pressures were integrated in the sense that, as in an organism, change in one affects another and always results in some compensatory regulation in the system. It seems to me that there is still considerable merit to the integrative scheme. Such a scheme admits to a tremendously complex interaction of factors and pressures that determine population density at a particular time and at a particular site.

It seems to me that compensatory responses have been largely neglected in most experimental investigations of population dynamics.

The kinds of responses envisioned are analogous to the compensatory responses that occur within the body of a single animal. Heart muscle hypertrophies in response to increased arterial pressure and to decreased renal glomerular filtration. When one kidney is removed the remaining one enlarges enough to compensate for the lost renal function. When nine-tenths of a rat's liver is removed, the animal survives and the liver tissue regenerates until the original mass is restored. When the hematopoietic tissue in bone marrow is insufficient to replace blood cells lost through chronic bleeding, extramedullary hematopoietic tissue appears in the spleen.

How can compensatory mechanisms be found at the population level? Drone elimination occurs in colonies of honeybees in the fall when the nectar flow diminishes. Workers herd the drones into corners of the hive and cut them off from the food supply. The weakened drones are ultimately dragged out and left to die outside the entrance of the hive. In one instance the exit to a hive was closed at the time of the drone slaughter. The workers cut up the bodies of the drones and dropped the remains through a crack in the bottom of the hive so that a conspicuous mound of drone parts accumulated directly below the hive. Queenless colonies may tolerate drones indefinitely. These two responses in a highly integrated social population were clearly compensatory. The way to discover compensatory mechanisms is to make a change in the population or in the enviroment and then wait to see what happens.

Exploitation Experiments with Sheep Blowfly

A.J. Nicholson[2] conducted two series of experiments in which he exploited populations of insects as a predator might do, then recorded the responses. Populations of sheep blowflys *(Lucilia cuprina)* were initially left free to develop and maintain themselves under predetermined environmental conditions for long periods, generally about a year. Then four populations were subjected to different levels of exploitation:

A — control undisturbed.

B — 50 percent of newly-emerged adults removed regularly.

C — 75 percent of newly-emerged adults removed regularly.

D — 90 percent of newly-emerged adults removed regularly.

Although the actual numbers in the populations varied greatly during the long term of the experiment because of a lag in the time it took for eggs to develop into adult recruits, the responses within each population always tended to restore the density toward a mean level.

Broadly, the result of destroying adults was to increase their rate of recruitment. The results were as follows:

A — $\bar{x}$ = 573 adults emerging daily.
B — $\bar{x}$ = 712 adults emerging daily.
C — $\bar{x}$ = 878 adults emerging daily.
D — $\bar{x}$ = 1,260 adults emerging daily.
A — $\bar{n}$ = 2,520.
B — $\bar{n}$ = 2,335.
C — $\bar{n}$ = considerably less than A.
D — $\bar{n}$ = considerably less than A.

As a compensatory response in populations C and D, average life span of the adults was prolonged from approximately 4.5 to 7 days and the adult recruitment rate much increased. What Nicholson called the minimum coefficient of replacement, taken as one in the control population, rose to 2, 4 and 10 in B, C, and D. This indicates an extraordinary degree of resilience to destructive pressure.

In these experiments there was a limiting factor, namely a restriction of 0.5 grams per day of protein in the otherwise unlimited food supplied the adult flies. Quantity of protein was previously known to be positively correlated with egg production. However, under experimental conditions population D with about one-third as many adults as population A produced an average of six times as many eggs per day.

In the second *Lucilia* experiment, the food supply for the developing larvae was limited uniformly in each of four populations E — H; the adults received a superabundance of all nutrients including protein. The populations were exploited as follows:

E — control undisturbed.
F — 75 percent of newly-emerged adults removed regularly.
G — 95 percent of newly-emerged adults removed regularly.
H — 99 percent of newly emerged adults removed regularly.

Although the average number of adults in each exploited population was kept down by this high mortality compared with the control population, all three populations were able to match adult losses with recruits from the developing larvae. The minimum coefficient of replacement for population H was 100 times the control value.

These experiments reveal several points of interest — for instance, an amazing propensity to build up the recruitment rate of the life-history stage immediately preceding the one subject to destruction, either eggs or pupae or both in these experiments.

Exploitation of Woodchuck Populations

Comprehensive study of woodchucks in south-central Pennsylvania patterned after Nicholson's experiments with confined insect populations demonstrated similar compensatory responses to exploitation[3],[4] One population was exploited by removing yearling and adult woodchucks at an average rate of 30 per month for eight months, February through September, of 1957 and 1958. The sex ratio of a second population was altered by removing 37 females in 1957 and 73 in 1958. An attempt was made to maintain a relatively undisturbed population in an adjacent reference area.

Woodchucks belong to the family Sciuridae of the order Rodentia. They have a combination of the fossorial and terrestrial adaptations and hibernate during the colder months of the year. Their principal foods are grasses and the leaves of clover, alfalfa, plantains and various perennials. Certain characteristics of these large rodents make them highly useful subjects for studies of population dynamics. They are relatively easy to trap, thus numbers can be determined by various routine methods of censusing. Females have no more than one litter of young each year, which in south-central Pennsylvania are born during April, thus populations go through annual cycles in which numbers increase abruptly after the young are born then decline gradually until the next breeding season. Reproductive performance can be determined by collecting and examining a representative sample of the females during the breeding season each year. At this time percentage pregnancy, reproductive rates and birth rates can be calculated by counting corpora lutea, viable and resorbing embryos and fetuses, and placental scars. Mortality rates can be calculated indirectly by trapping, marking, and releasing a number of animals in a population one year and then determining how many disappear by the following year. The extent of movements can be determined by noting the locations at which marked animals are trapped. Thus, each of the three forces of population growth can be measured relative to population size and composition.

This study of woodchucks was conducted on the Letterkenny Army Ordnance Depot which covers nearly 20,000 acres. The investigation was limited to about half of this area, where topography and ground vegetation were highly uniform. Here woodchucks were at least tenfold more numerous than in the intensively cultivated farmlands adjacent, at times approaching densities of one animal per acre. This section of the depot is divided into 15 rectangular units of 300 to 600 acres which are separated by zones of cultivated fields 1,300 to 1,500 feet wide. Three of these units (see Figure 26) were selected as study areas: area C, 600 acres, the exploited study unit; area D, 535 acres, the reference study area; and area G, 333 acres,

where the sex ratio was changed. The zones of cultivated fields
between these smaller study areas, although not barriers to movement,
separated the populations geographically. Plowing and cultivation
in the buffer zones discouraged woodchucks from establishing
permanent burrow systems. The food supply was always in excess of
the needs of the woodchucks, as from 90 to 95 percent of the land
consisted of meadows of grasses (especially Kentucky bluegrass,
Poa pratensis), clovers, plantains and perennial weeds.

Figure 26. Woodchuck Movements in Three Pennsylvania Study Areas.
  Numbers of Emigrants and Immigrants are Indicated in Vectors.

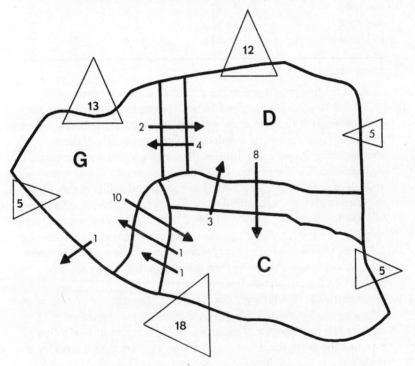

The three populations were of equal density in the spring of 1957
at the beginning of the study. Although more than 500 woodchucks
were removed from area C in a two year period (about five times as
many as were taken from reference area D), censuses each spring and
again each fall in 1957 and in 1958 showed no significant differences
in the densities of populations C and D. Table 26 indicates the numbers

of animals caught with the same trapping effort in each study area. Area G was not trapped until 1959 because the population level there was not pertinent to the studies until 1959 when the sex ratio had been altered significantly from the original one to one ratio.

Table 26. Number of Animals Caught with the Same Trapping Effort on Each Study Area[3]

Period	Area C	Area D	Area G
Sept. 1957	41	40	*
Mar. 1958	67	68	*
Aug. 1958	62	63	*
May 1959	51	41	29

* No Trapping

The breeding population in the protected study area each spring contained 70 percent adults and 30 percent yearlings (young born the previous year). This population was relatively stable; that is, it contained about the same number of woodchucks each spring. Under these circumstances 30 per cent of the adults died each year and were replaced by reproduction. The exploited population contained just as many animals per acre each spring but the proportion of adults in this population each spring changed from 70 percent in 1957 to 57 percent in 1958 to 39 percent in 1959 (Table 27). A number of compensatory responses were initiated that prevented an overall reducation in density following the removal of so many woodchucks. More young were born, more young animals surviving than heretofore, and young animals moved into the population from the exterior.

Of particular interest were the changes in reproductive performance following exploitation (Table 28). Although the mean number of corpora lutea per pregnant female, which is a measure of the ovulation rate, and the mean number of implantations (an embryo, a fetus, or a placental scar is evidence of an implantation) per pregnant female declined from 1957 to 1959, the birth rate was highest in 1958. The ovulation rate, average litter size, etc., increases with parity in woodchucks, thus the decline in ovulations and implantations must have been a reflection of a lower mean age of the females, as this population contained proportionately more yearling and fewer adult females each year. Prenatal losses were less following the removal of adult woodchucks from this population; hence, the average litter size

Table 27. Improvement in Reproduction as Woodchucks were Removed from Area C[4]

| Means per Pregnant Female | Status of Population | | | | | |
| | 1957 | | 1958 | | 1959 | |
	Dense		Removal of animals		Removal continued	
Mean Corpora Lutea	5.46	(71)	5.27	(52)	4.71	(31)
Mean Implantations	4.67	(76)	4.34	(83)	4.03	(21)
Mean Embryo Loss	−1.25	(32)	−0.23	(31)	−0.39	(23)
Mean Litter Size	3.42		4.11		3.64	
Total % ♀♀ Pregnant	75	(101)	79	(106)	61	(51)
% Yearlings in Population	30	(122)	43	(124)	61	(161)
% Yearlings ♀♀ Pregnant	20	(31)	56	(50)	42	(33)
% Adult ♀♀ Pregnant	100	(70)	100	(56)	94	(18)
Birth Rate	1.29		1.63		1.11	

*Numbers of animals examined to obtain each value are given in parentheses.

Table 28. Reproduction in Areas D and G[3]

| Means per Pregnant Female | Area D | | | | Area G | | | |
	1958		1959		1958		1959	
Mean Corpora Lutea	5.39	(33)	4.61	(23)	5.23	(13)	4.46	(24)
Mean Implantations	4.86	(36)	3.91	(24)	4.76	(17)	4.00	(24)
Mean Embryo Loss	0.66	(32)	0.29	(21)	*		1.19	(21)
Mean Litter Size	4.20		3.62		*		2.81	
Total % Pregnant	84	(50)	82	(28)	71	(17)	61	(51)
% Yearlings in Population	30	(122)	28	(102)	25	(28)	44	(110)
% Yearlings Pregnant	43	(14)	20	(5)	43	(7)	20	(20)
% Adults Pregnant	100	(36)	96	(23)	90	(10)	87	(31)
Birth Rate	1.76		1.49		*		0.51	

*Insufficient data.
( ) Numbers of animals examined to obtain each value are given in parentheses.

(or the average number of live young produced per pregnant female) was higher in 1958 than in 1957. There was a definite improvement in reproduction because a higher proportion of the yearling females became pregnant and intrauterine mortality of embryos and fetuses was considerably less. An increase in the percentage of yearling females pregnant from 20 per cent in 1957 to 56 per cent in 1958 and 42 per cent in 1959 is especially significant because experiments with confined populations of mice show a delay in the attainment of sexual maturity caused by crowding. Remember also that crowding increased intrauterine mortality in laboratory mice.

Reproduction was lowest in area C in 1959, but then only 40 per cent of the breeding females were more than a year old and apparently no more than 50 to 60 per cent of the yearling females become pregnant even when social pressures are reduced. A point must have been reached when the average age of the females was too low to allow maximum production. Table 28 contains reproductive data for areas D and G in 1958 and 1959 for comparison.

Reproduction was also improved in the reference area by 1959 but not to the same extent as in the exploited area. This was associated with an overall decline in density in the two areas (see Table 26). The summation of density-dependent responses, especially the movement of young woodchucks from densely to less densely populated places caused numbers to level out over relatively large areas. The removal of animals from one locale had far-reaching consequences and affected densities in surrounding areas as well.

The removal of females from area G in 1957 and 1959 produced a sex ratio in the spring of 1959 of approximately 58 percent males. Population density was substantially the same or even below that in the two adjoining study areas (see Table 26). Reproductive rate was well below that in the adjacent study areas, 1.20 compared with 2.21 in the exploited area and 2.97 in the reference area. Poor productivity resulted because of failure of some adult females to become pregnant (an unusual occurrence), a low incidence of pregnancy of yearling females (20 percent), and a high incidence of embryonic and fetal resorption (1.19) per pregnancy. The reproductive results in area G suggest that population pressures, in this instance the rate of aggressive interaction between males, was increased by altering the sex ratio to produce a preponderance of males.

The numbers of woodchucks marked and released in areas C and D and the proportion recovered subsequently a year later are listed in Table 29. Animals are grouped according to age. Those marked as yearlings or adults are referred to as old woodchucks, while those first marked as young of the year are categorized as young woodchucks.

Table 29. The Proportion of Marked Woodchucks that were Recovered during the Year after they were Released[3]

	Exploited Population		Protected Population			
	(C)		(D)		(G)	
	Old	Young	Old	Young	Old	Young
Marked and released one year	48	154	159	101	80	114
Recovered the next	20	50	85	23	42	30
% recovered	42	33	54	23	53	26

More than 50 percent of the old woodchucks marked and released in the area from which few animals were removed compared with less than 25 percent of the young of the year were subsequently recovered. In contrast the proportion of young recovered from the exploited population was not significantly different from the proportion of older animals recovered from this population. I have already presented evidence that normally the mortality rate is highest among the young. The significant point is that young woodchucks survived as well or nearly as well as the older woodchucks in the population after a portion of the older animals had been removed. Increased survival of the young animals to approximate that of the older animals is clearly a second compensation for a reduction in density.

The distances woodchucks moved between successive points of capture were recorded during this study to determine if movements were related in any way to population density. The young of the year, as expected, moved over a much larger area than the adult woodchucks; also a portion of the young in each study area moved away or emigrated across the boundary zones into adjacent areas. Significantly more young woodchucks emigrated from areas D and G than from the exploited area, which would indicate that the movement of young woodchucks was influenced by population pressures (see Fig. 27). Only four woodchucks marked and released initially (1.9 percent of 154, one was captured in the adjacent buffer zone) in the exploited area were recaptured subsequently in other study areas, while 18 woodchucks marked initially (8.4 percent of 215) outside of this area were subsequently recovered within its boundaries. Apparently young woodchucks emigrated from dense populations on the exterior and settled in the area from which so many adult woodchucks had been removed. The implication is not that movements were directed to this area because the population density was low, but rather that young woodchucks moved into the area at random and remained there because population pressures were low. A tendency for young animals to move into the population as the older animals were eliminated would be a third compensation for a reduction in density.

A summary of the data on movement is presented in Tables 30 and 31. An analysis of these data shows the young animals, especially the males in area D and the females in area G, averaging much greater distances between captures than the older woodchucks. Actually young of the year traveled distances from 3,000 to 10,000 feet between successive points of capture. Proportionately more males than females emigrated from area D, where the population contained only slightly more females than males, while proportionately more females than males emigrated from area G, where the population contained a significant preponderance of males. Nearly the same proportion of young emigrated from each area, but the sex ratios of the emigrants were approximately reversed. These results suggest that the movement of young woodchucks was influenced by the sexual structure of the population as well as by the population density. Consequently, emigration must play an important role in the dynamics of these populations. The young males may be more sensitive to population pressures, as they traveled greater distances than young females generally. Greater movement of young females in area G was a surprising event. Such movements would not compensate for the altered sex ratio, but actually increase the disparity among the sexes.

The results of this study illustrate the principle that natality, mortality, and movement, as forces of population growth, respond to changes in density. These are truly compensatory responses that tend to adjust the population level to the limits set by the carrying capacity. In areas that were maintained as dense populations only 20 percent of the yearling females became pregnant and intrauterine mortality was high. Few young woodchucks survived to maturity and many emigrated from these populations. Clearly these responses would limit population growth. On the other hand, following the elimination of a portion of the older animals from an area, reproduction improved, more young woodchucks survived and remained in the area, and at the same time several young animals moved into the population. Such responses would promote population growth.

Table 30. Mean Distance in Feet between Successive Captures. (Number of movements given in parentheses)[3]

Area	Old ♂♂	Old ♀♀	Young ♂♂	Young ♀♀
C	490 (43)	373 (44)	811 (62)	682 (43)
D	465 (108)	372 (149)	1,845 (29)	623 (33)
G	390 (105)	421 (81)	928 (65)	1,710 (26)

Table 31. Percent of Movements Over 3,000 Feet. (Number of Movements Given in Parentheses)[3]

Area	Old ♂♂	Old ♀♀	Young ♂♂	Young ♀♀
C	2.3 (44)	0.0 (44)	4.8 (62)	7.0 (43)
D	0.9 (109)	0.0 (149)	24.1 (29)	6.0 (33)
G	0.0 (105)	0.0 (81)	7.7 (65)	19.2 (26)

The summation of these density-dependent responses would cause densities to level out over relatively large areas, which explains why it was impossible to maintain different levels of densities in areas not widely separated. Age compositions were not the same in the three areas because of the tendency for the older animals to remain in a home area once they were established there.

The data on movement of woodchucks give perhaps the most important single clue to the role of social factors in regulating population growth. Records from this study include 790 movements of woodchucks from one trap site to another. The distance traveled between successive points of capture for adult woodchucks averaged 410 feet in contrast to the young of the year which after weaning moved an average of 1,000 feet from one point of capture to another. Young woodchucks were apparently moving about in search of a home area, home region, or whatever term is applicable. On the other hand, the relative lack of movement on the part of the older animals is basic to the concept that social pressures increase in a growing population. Once a woodchuck becomes established in a certain location, it apparently remains there for long periods of time. Trapping records show that woodchucks have stayed in one place, a home area, for as long as three years. The penchant for establishing home ranges or territories is significant, I think, because there must be several social or behavioral interactions involved to allow such an arrangement. Animals that do not succeed in establishing a home area apparently do not survive, since there were no records of extensive movements once an animal was more than a year old. Home areas are apparently maintained by a sort of mutual tolerance on the part of neighbors in the population. Increased density must increase social pressures making it more difficult for neighbors to maintain their social position toward one another. Young animals, once they are weaned and leave the maternal den, also must find a place in this social structure. When competition for social position is intensified neuroendocrine mechanisms that reduce productivity and increase mortality are activated. The subordinate members of the population are undoubtedly the most affected by increased social competition.

Compensatory Responses to Altered Sex Ratios

An unusual sex ratio was found in the young of the year trapped in area G in 1958 (40 males and 89 females). Areas C and D, as well as area G in 1957 before females were eliminated, had populations with normal or essentially unaltered sex ratios. The sex ratio of young woodchucks captured in these populations was 211 males to 204 females. The difference between the sex ratio of the young trapped in area G in 1958 and that of the young trapped in the reference and exploited populations was highly significant statistically ($P < 0.005$), thus the occurrence could not be dismissed as a fluke of sampling. Moreover, this unusual surplus of females was more than adequate compensation for the shortage of females created. Of course the elimination of females from area G continued in 1958 to create the desired alteration in the breeding sex ratio, but the fact remains that a satisfactory compensatory response for the disparity in the sexes had already occurred.

These results were similar to those reported by White (1914) in a population of rats in India. He described a change in the sex ratio of newborn rats associated with unusual mortality of adult females during an epidemic of plague. The following is quoted from White's paper:[5]

As if to compensate for the apparently wholesale destruction of adult females, females only appear to have been born. These two processes, the destruction of females and the suppression of male births proceeded *pari passu*. In June not a single male rate below the weight of 80 grams was trapped, whereas 610 females of less weight

than 30 grams were caught. As the number of adults of the two
sexes began to approximate more closely the one to the other, young
rats were again trapped in increasing number.
No explanation was offered for this phenomenon.

Apparently the sex ratio of the young trapped in area G was a
reflection of the sex ratio of young born to females in this area in the
spring of 1958. As dispersal of the young does not commence until the
beginning of July, the sex ratio of the young trapped during May or
June should reflect the sex ratio of young produced in the area. Only
25 percent of 59 young woodchucks caught between 15 May and
3 July were males. Thirty-five percent of the 70 young woodchucks
captured between 4 July and 15 August were males. More young
females than young males emigrated from area G; also the five
immigrants captured and identified in area G were all males. Hence the
movement of young females into area G or the emigration of young
males from area G was discounted.

Young females were not easier to trap than young males at any time
of the year. Fifty-one percent of the 102 young of the year captured
in area C and 57 percent of the 51 young captured in area D between
15 May and 3 July in 1958 were males.

The sample of 129 young woodchucks caught on the 333-acre study
area represented a very large proportion of the total number of young
produced in this area in 1958. Unfortunately, due to the short breeding
season of woodchucks and the high incidence of embryonic and fetal
resorption encountered in area G, only seven females carrying fetuses
old enough to determine fetal sex were collected in 1959. The sex ratio
of this small sample of fetuses favored the females by nearly two to one
over the males ($10\male$: $17\ \female$). If a high proportion of neonatal losses
were males, the sex ratio of young woodchucks after weaning would
have favored the females. However, it is difficult to visualize factors
that would affect the sexes differentially between birth and weaning.

After all possibilities were considered it was apparent that more
females than males were born in area G in 1959. Assuming no vagaries
of sampling, there were three possible biological causes of the deviation
in sex ratio: (a) higher fetal mortality for the males, (b) the X
chromosome-bearing sperm was favored at conception, and (c) sex
reversal, genetic males developing as somatic females.

Any number of factors, maternal age, paternal age, parity, season,
birth order, litter size and so forth has been reported in the literature
as capable of altering the sex ratio. However, in the present study none
of these factors appeared of any importance. The only common

denominator in the populations of wild rats in India and wild woodchucks in Pennsylvania was a high level of stress. The female rats were afflicted with disease and the female woodchucks, judging from their poor reproductive performance, were also responding to stress.

Explanations for these strange compensatory responses to alterations of the sex ratio of adult breeders are completely hypothetical at this point. On the basis of genetic theory, spermatozoa determining the male and the female sex are produced in equal numbers. However, many factors might differentially affect the spermatozoa between their formation and the fertilization of ova. Stress of social competition affects both male and female reproductive tracts. Therefore it is possible that hormonal changes that affect the male reproductive tract directly also cause differential mortality of spermatozoa. There is experimental evidence that this is the case in brown house mice.[6] Endocrine responses to stress can also affect the uterine environment, thus increasing intrauterine mortality or favouring the development and survival of one sex over the other. The last possibility is suggested in two recent papers. The rate of reduction of triphenyltetrazolium to formazan was used as an index of estrogenic effects, and examined in relation to reproductive performance.[7] Female rats with high uterine reaction rates produced litters with a higher percentage of female pups than females with a low level of reaction. In this study, litter size was not related to uterine metabolic activity. Modification of the sex ratio in rats was accomplished by the administration of ACTH (the stress hormone) to mothers during early pregnancy.[8] The number of females in the litters was increased by about 10 percent over that in control litters.

Changes in sex ratio of the young woodchucks in this study were compensatory, but again the responses may have been fortuitous. Altering the sex ratio to proportionately lower the number of males in a population normally decreases social pressures. The primary sex ratio in the absence of stress factors seems to favor the male sex.[3,6,7,8] Thus, the compensatory responses with respect to sex ratio seem to be operating in the correct manner. Whether such mechanisms would readjust population sex ratios in nature is an interesting possibility. An expenditure of effort toward investigating such compensatory changes would seem justified in view of these intriguing studies on the subject.

Summary and Conclusions

Compensatory responses have been largely neglected in most experimental investigations of population dynamics. Competition for the necessities of life appears to be the underlying and fundamental driving force of plants and animals. Thus, populations as collections of interacting individuals sharing common genes (a Mendelian population) are complex study units which enable biologists to assess the total impact of a species within the community structure or the ecosystem. At times the individuals act in concert, co-operate, flock, herd, or form organized social groups. Such behavior while obviously being advantageous to each individual enhances the survival of the group. Whether the advantage collectively is only a fortuitous result of the advantage to the individual will probably always be a nagging question. On the other side of the coin competitive interaction at times involves hostile behavior, mutual intolerance, vindictiveness, hoarding and the like. From either standpoint the unsympathetic observer might say the individual is always looking out for number one.

The concept of compensatory interaction however needs no theoretical deductions. It describes the many different ways individuals or groups can compensate for changes that occur either in the physical environment or within the biotic structure of the group, population, or community. Animals are often given credit for much ingenuity. Thus, great horned owls nest on cliffs when tall trees are scarce and Canada geese will take advantage of platforms placed on high poles or trees for nesting sites. Ruffed grouse drum on boulders in second growth forests where fallen tree trunks have decayed into powder. Thus, success of populations in my opinion should be measured in terms of their degree of adaptability. Humans take much credit for being uniquely adaptable to a wide range of environments. Sociologists are inclined to attribute such adaptability to the high degree of development of the human brain. Implications of the research on animal populations will be discussed in the next and last chapter.

Notes

1. J.B. Calhoun, 1950. 'The study of wild animals under controlled conditions'. *Annals N.Y. Acad. Sci.* 51: pp.1113-22.
2. A.J. Nicholson, 1955. 'Compensatory reactions of populations to stresses, and their evolutionary significance'. *Aust. J. Zool.* 2: pp.1-8.
3. R.L. Snyder, 1960. 'Physiologic and behavioral responses to an altered sex ratio of adults in a population of woodchucks,' Doctoral dissertation, Johns Hopkins School of Hygiene and Public Health, Baltimore, Md.
4. ———— 1961. 'Evolution and integration of mechanisms that regulate population growth'. *Proc. Nat'l Acad. Sci.* 47: pp.449-55.
5. F.N. White, 1914. 'Variation in the sex ratio of *Mus rattus* associated with an unusual mortality of adult females'. *Proc. Roy. Soc. London.*87: pp.335-44.

6. R.L. Snyder, 1966. 'Fertility and reproductive performance of grouped male mice'. In *Comparative Aspects of Reproductive Failure,* K. Benirschke, ed., pp.458-72. Springer-Verlag, New York, Inc.
7. A.B. Schultze, 1965. 'Litter size and proportion of females in the offspring of multiparous rats with varying uterine metabolic levels'. *J. Reprod. Fertil.* 10: pp.145-7.
8. E. Geiringer, 1961. 'Effect of ACTH on sex ratio of the albino rat'. *Proc. Soc. Exp. Biol. Med.* 106: pp.752-4.

8 IMPLICATIONS TO HUMAN POPULATIONS

Planet Earth came into being some two or three billion years ago probably as the consequence of a 'near collision' of the sun with another large stellar body. The origin of life on earth goes back approximately one billion years. The earth's crust has retained a record of the evolution of living beings for approximately half-a-billion years. The gradual change from inanimate carbon compounds in colloidal systems to protoplasmic material complete with protective membrane occupied the prior 500 million years. Man appeared on the scene barely two million years ago, which accounts for only two-tenths of one percent of the entire span of evolution. Undeniably modern man *(Homo sapiens)* has made more of an impact on the physical and biotic environments of the earth than any other species in recorded history. There has never been anything like it, thus the ecologists among populations of men have been busy analyzing the results of the interactions of their fellows with the earth's environment and other creatures.

Practically the same questions that have been posed for infrahuman populations are being asked about human populations. Is *Homo sapiens* a successful species? In the general scheme of evolution is man progressing or regressing? What are the prospects for the future?

The answers to any of these questions require value judgments. If success is measured in terms of competition with other species of plants and animals for the planet's resources, *Homo sapiens* is at least holding his own. Small rodents and certain insects have been competing extremely well; in some cases they have gained the upper hand in exploiting food and shelter. Man has been better able to deal with those creatures of equal or larger size; the fates of the American bison and the mountain lion are good examples of the outcome when man decides to compete with another species for food (prey) or space. The great whales are another matter. They are simply being exploited without regard to population fundamentals.

Possibly there are persons who care nothing about other species on earth. Such persons might decree that domesticated plants and animals specially bred to provide the highest sustained yields of essential nutrients and materials should occupy all of the available spaces on earth. On this kind of earth, man would reign supreme at the ends of all food chains. At the same time, there are those persons on earth who contend that such an arrangement would create a living hell, that man's psychological well-being is intricately bound up in wild things, and that somehow we should share this earth with other life forms. I suspect

211

what any one individual would like to see is based on early experience and education as much as anything else. Thus, my opinion of what are beneficial consequences of man's activities should be considered for what it is — a value judgment.

Lately, man's harmful activities with respect to other species have been receiving most of the publicity, so I will cover the endangered species first. There seems to have been a pattern of human behavior repeated over and over again since man turned from hunting and food-gathering to animal husbandry and agronomy. While searching the mountain chains of south-central Pennsylvania for wild turkey flocks in 1953, I was surprised to find so many old barn and farmhouse foundations in the second growth forests of the foothills and mountainsides. The remnants of an old canal system along the Juniata River nearby were equally intriguing. I finally realized that the land around Harrisburg, Pennsylvania had been cultivated much more extensively about 150 years ago. The ruins of farm buildings and canals represented a way of life before the advent of the steam engine. The fertile valleys and flatlands between the mountains are still occupied by prosperous farm families. The foothills and mountainsides are badly eroded and large areas that once provided pasture for large numbers of cattle and horses support only sparse populations of grasses, shrubs, and small stunted trees today.

I can only conjecture about life in the last century. Horses and oxen provided the primary modes of transportation before steam locomotives and gasoline-driven motor vehicles came on the scene. Since beasts of burden required a good supply of grains and forage, farmers were able to make a suitable living on marginal farmlands. Actually, the times demanded the exploitation of all available lands because the slow forms of transportation required that farms be located in close proximity to the centers of population. Steam locomotives changed everything and also opened up vast areas of fertile lands in the middle and western portions of the United States. Many farmers were still cultivating the shallow, less fertile soils along the mountainsides at the turn of the present century, but most of those remaining were relocated during the recovery stages of the Great Depression in the 1930s.

It would be interesting to speculate about the men and women who moved westward as population growth continued in the New World. I would imagine that the first-born inherited the farm and that younger sons and daughters were forced by necessity to join the emigrating portion of the population. The original emigrants to the New World were probably the younger, less fortunate ones who were more the

victims of circumstances than anything else. The ancestral farm in
Europe or the British Isles may have already been subdivided several
times to provide for the sons and grandsons of the original landowners,
but like the mice at asymptotic population densities, the point was
finally reached when the youngest son had to look for new horizons.
The principle outlined in Chapter 5 with respect to the temporal
circumstances of birth undoubtedly applies to human populations as
well.

The characteristic pattern of land abuse is only too well known and
continues today. Devastating erosion is the price of misuse of land. It
begins when the trees of the climax forest are cut. The exposed topsoil
is washed away gradually and the process is hastened by plow and
cultivator. The practice of setting fires to clear grasslands of undesirable
'weeds', a widespread and supposedly beneficial technique, brings
further ruin. Seedlings of trees and shrubs that might have saved the
soil are destroyed. Grass roots are burned. When the earth is totally bare
it is baked by the sun into lifeless brick. Rain cannot penetrate the soil
but forms torrents that undermine hillsides causing furrows and gullies.
Thus, once-fertile productive land becomes forever useless. Vast areas
of the Middle East and Northern Africa have undergone just such a
metamorphosis to climax in barren deserts.

Mismanagement of land in the scheme of things cannot be stressed
too much. A shortage of animals brought on by over-harvesting can often
be remedied, but destruction of the habitat cancels even the chances of
reform in an enlightened future. Soils washed or blown to the bottoms
of the oceans cannot grow crops to feed hungry animals, human or
otherwise.

Pollution of the environment with chemical and solid by-products
of civilized man, especially when the environment consists of fresh
water lakes and streams, can wipe out entire species, food chains, and
ecosystems. The estuaries of the Thames, Humber and Tees all supported
important shrimp, eel, flounder and smelt fisheries at one time, and
some had mussel and oyster beds. Even twenty years ago an inshore
vessel could catch 1,000 bushels of shrimps in a single haul. The
Directors of Fisheries Research at Lowestoft, England, points out that
we are becoming so accustomed to their absence that we forget these
resources once existed.

The number of endangered plant and animal species around the
world is so staggering that it is difficult even to summarize the situation
adequately. The Survival Service Commission (SSC) of the International
Union for Conservation of Nature and Natural Resources (IUCN) has

published 'Red Data Books' (RDB) on mammals, birds, amphibians, reptiles, freshwater fish and angiosperms. The first RDBs issued in 1966 compiled information on all threatened species. Hopefully revisions of the Red Data Books are keeping everything up to date; the fact that several volumes are required to cover the subject of endangered species is a measure of the calamity.

Perhaps what has happened to wildlife species in the last 100 years or so can best be appreciated through reading historical accounts of one's own home territory. Animals extirpated from Pennsylvania in the recent past include wood bison *(Bison bison athabascae)*, moose, wapiti *(Cervus elaphus canadensis)*, timber wolf, mountain lion, beaver, marten, fisher, heath hen *(Tympanuchus cupido)*, passenger pigeon, and sturgeon *(Acipenser sp.)*, American shad *(Alosa species)*, impoitant and valuable commercial fishes since the earliest settlements in North America, were blocked permanently from the Susquehanna River by power dams in this century and for a considerable period of time from the Delaware River by organic wastes dumped into the lower river and Delaware Bay.

Now, let us consider the other side of the coin — successful conservation efforts and reintroduction of animals into Pennsylvania habitats. The wood bison has not returned to Pennsylvania. A few members of the species survive in a remote corner of the Wood Buffalo Park between Great Slave Lake and Lake Athabasca which was established by the Canadian government in 1915. For a time the continued existence of this subspecies was in doubt because thousands of surplus plains bison were also brought there from another preserve. These animals spread tuberculosis and crossbred with the wood bison. Fortunately, a last herd with approximately 200 pure wood bison was discovered in 1960 so there is still a chance that the subspecies will be preserved.

The moose could conceivably live again in Pennsylvania but to my knowledge no one has attempted a restocking program. The wapiti was reintroduced in 1913 when 50 were brought in from Yellowstone National Park and 22 were purchased from a local game park. The new arrivals were from an alien race, *Cervus elaphus nelsoni*. The original eastern subspecies had long since disappeared from the North American Continent; the last native wapiti was probably killed in Pennsylvania in 1867. By a strange quirk of fate the *canadensis* gene pool was saved from total extinction because animals of the eastern race had already been introduced to New Zealand. The genes survive in the red deer population of that country.

The rocky mountain wapiti survives today in the rugged mountains of north-central Pennsylvania. The last census five years ago set the wild population at 80 ± 10 individuals. An open hunting season ran from 1923 to 1931 during which 98 wapiti were killed legally. Wapiti have been protected by closed hunting seasons since 1931.

Timber wolves could also be restocked in their native habitats in Pennsylvania, but the objections of the state's human residents would be a large stumbling block to success. The wolf species has a bad reputation, most of it unjustified. The same is true of the mountain lion. The eastern subspecies of this graceful felid may still exist in New Brunswick, Canada and there are fanciful accounts of mysterious large cats inhabitating remote regions of Pennsylvania. A mountain lion was actually killed by a hunter in north-western Pennsylvania recently, but this animal was almost certainly a specimen escaped from captivity or deliberately released by someone.

The reintroduction of the beaver to Pennsylvania is a success story of major proportions. Beaver were practically exterminated about 1830. The last stragglers were killed between 1850 and 1865. Beaver killed in 1884 and one seen in 1899 were probably animals that had escaped from captivity. A pair was imported from Wisconsin in 1917. From then until 1924, 94 more beavers were imported from various spots by the Pennsylvania Game Commission at a cost of about $50 each. In 1931 conservation officers counted 899 beaver dams and estimated the population in the state at 4,377 animals. In 1934 the count was 15,000 with 6,455 legally trapped for their pelts.

Marten and fisher are not extinct and may eventually extend their ranges again to include Pennsylvania. Unfortunately the heath hen and the passenger pigeon are gone forever.

The Pennsylvania Game Commission deserves most of the credit for preserving the native wildlife and replacing extirpated species. There are a few men and women among every group of hunters and fishermen who are ardent conservationists. Some would say they act from selfish reasons, but I find that hard to accept. All of us join in the food chain at one level or another. If we eat no meat we still compete with other species for green vegetation. Hunting and food gathering, once the primary techniques for obtaining food, have only been sophisticated with the inventions of tractors, combines, and feedlots. We can hardly escape the role that survival dictates. Thus, the hunting and fishing conservationists were always in the forefront. The first employees of the Pennsylvania Game Commission are especially worthy of praise because they were attempting to protect wildlife when many of our citizens were still convinced that animal populations would last forever.

In the same vein one might ask what hope for the future? There is one simple answer in my opinion. The human animal will permit other animals to share the environment if there is profit in it. For example, wildlife preserves for the larger ungulates and their predators in Africa may be expected to remain if their economic return to the native inhabitants is greater than the income that could be derived from farming and ranching. National parks in the United States and Canada will remain as long as visitors spend money during their travels. When coal, oil, and minerals lying beneath the ground in many wildlife preserves are considered more valuable than the animals and their habitats, the latter may suffer the consequences. When the value of one resource outweighs the other, the governments of men nearly always select the resource of higher worth. This problem was mentioned in the beginning of the first chapter, but it is worth repeating. 'Assessment of environmental impact requires certain judgments about the values of the organisms affected.' Unfortunately, values at present are based almost entirely on monetary considerations.

In the final analysis the fundamental laws of population are concerned with numbers. Thus, the population density of the dominant species, *Homo sapiens,* is the overriding influence. Assuming that the first representative of the species appeared about 2 million years ago, we can estimate that between 60 and 100 billion men and women have lived on Earth.[1,2] Today some 3.7 billion people inhabit the planet, roughly 4–5 percent of all those who have ever lived.[3]

Substantial historical data on which to base estimates of human populations are not available before 1650. Agriculture was apparently unknown before about 8,000 B.C. Based on the population densities of the hunting and food gathering tribes of today the total human population of 8,000 B.C. has been calculated at about 5 million people. The total human population at the time of Christ has been estimated at 200 to 300 million people; it had increased to about 500 million by 1650. Doubling times after 1650 are noted in Table 32. The population is now close to 3.9 billion and the United Nations forecasts five billion in slightly more than ten years at the present rate of growth.

With a world population of five billion and a doubling time of 35 or 40 years, it is evident that *Homo sapiens* is in the exponential part of its population growth curve. The principles of population growth dictate a leveling off phase or a severe decline in numbers, for one of the 'laws' of population states that no population can grow indefinitely. The 'law of the minimum' states that population growth

Table 32. Doubling Times of Human Populations After 1650 AD[1,4,5,6]

Date	Estimated World Population	Time for Population to Double
8000 BC	5 million	
		1,500 years
1650 AD	500 million	
		200 years
1850 AD	1,000 million	
		80 years
1930 AD	2,000 million	
		45 years
1975 AD	4,000 million	

will be limited ultimately by the factor in least amount.

Malthus expounded the thesis that human populations tend to increase up to the limits of the means of subsistence. Historically, it would be hard to prove that human populations have ever been truly limited in any one locale. Bubonic plague epidemics produced transitory sharp declines in the European population in the fourteenth and seventeenth centuries.[5,6] Ironically, although the 'Black Death' may have killed between 25 and 50 percent of the inhabitants of Europe and England at times, the overall growth curve was not altered. Down through history, war, famine, and pestilence have resulted in great loss of life, but compensatory, demographic forces have quickly brought the populations back to the base curve just as increased reproduction and immigration followed exploitation of woodchuck populations.

Actually it is incorrect to think of the world population as a single Mendelian population. There are instead subunits that fit the precise definition of population given in Chapter 1. These are collectively groups of people of one nationality bound in a frame of space and time. Immigration laws and social customs are powerful deterrents to intermixing populations. Therefore, several populations, notably those in the United States, most European nations, and Japan are beginning to approach stability or zero population growth (ZPG). Such populations are not being limited by environmental forces *per se* but by social customs that are reducing birth rate. In other parts of the world populations are exceeding or close to exceeding habitat-carrying capacity. In many countries food production could be increased by improvements in animal husbandry and agronomy. Many of the farming practices in the underdeveloped nations are self-defeating but introduction of better techniques for food production is difficult and expensive. A persistent scarcity of fertilizer is a problem because of the colossal size of the capital investment needed to build the necessary plants — 8 to 10 billion dollars each year. A limitation in the amount of land available for farming and ranching may eventually be a more serious problem. Finally there is the matter of water shortages. Many hydrologists believe that shortages of water, more than land, may ultimately limit food production for humans.

Summary and Conclusions

In general, population biologists are convinced that laws and principles governing population dynamics of plants and animals pertain to all living organisms including man. The major difference between man and other organisms relates to the brain. The power of speech and with it the ability to accumulate information from generation to generation are responsible for man's unique position on earth. Then, it should be possible for governments of men and men acting individually and collectively to control their own destinies. In other words, planning and corrective (compensatory) action can substitute for natural regulation. Natural regulation of animal populations is wasteful of flesh and individuality. Wise land use, wise resource management, and wise population control are the alternatives to natural regulation.

The primary problem besetting man today is uncontrolled population growth. This should be all too obvious by now. Until this problem is completely solved by reason, education, and example, it will be necessary to improve methods for food production and to reform technology to reduce waste and establish efficient operations. One way to achieve this ideal would be to turn to ecosystem management, which would place human populations in proper alignment with other animal populations in the natural system. In this way man might exploit all food chains on a maximum sustained yield basis. One outcome of ecosystem management may be a pleasant surprise. We might find that we can afford to allow many more species to exist at higher population levels in managed ecosystems than were ever dreamed of. Salmon, shrimps, shad, sturgeons, and eels all make fine meals for hungry people. Rivers, streams, lakes, estuaries, and oceans free of polluting chemicals and suffocating silt would increase the carrying capacity of planet Earth immeasurably. A reversal of soil erosion in the terrestrial habitats would be an added bonus. The science and technology required for ecosystem management is found in ecology. And understanding population is the key to success.

Notes

1. N. Keyfitz, 1966. 'How many people have ever lived on Earth?' *Demography,* 3: pp.581-2.
2. Population Reference Bureau, 1962. 'How many people have ever lived on Earth?' *Population Bulletin,* 18 (1)
3. United Nations, Statistical Office, 1974. *Demographic Yearbook.*
4. P.R. Ehrlich and A.H. Ehrlich, 1972. *Population, Resources, Environment: Issues in Human Ecology,* 2nd ed., W.H. Freeman and Company, San Francisco.
5. _____ , J.P. Holdren and R.W. Holm, (eds.) 1971. *Man and the Ecosphere.* W.H. Freeman and Company, San Francisco.
6. W.L. Langer, 1958. 'The next assignment'. *Am. Hist. Rev.* 63: pp.283-305.

INDEX